More BorderLines

30 Reproducible Borders

Illustrated by **Jean Thornley**

ISBN 0-931724-71-6

NOTE
NOTE

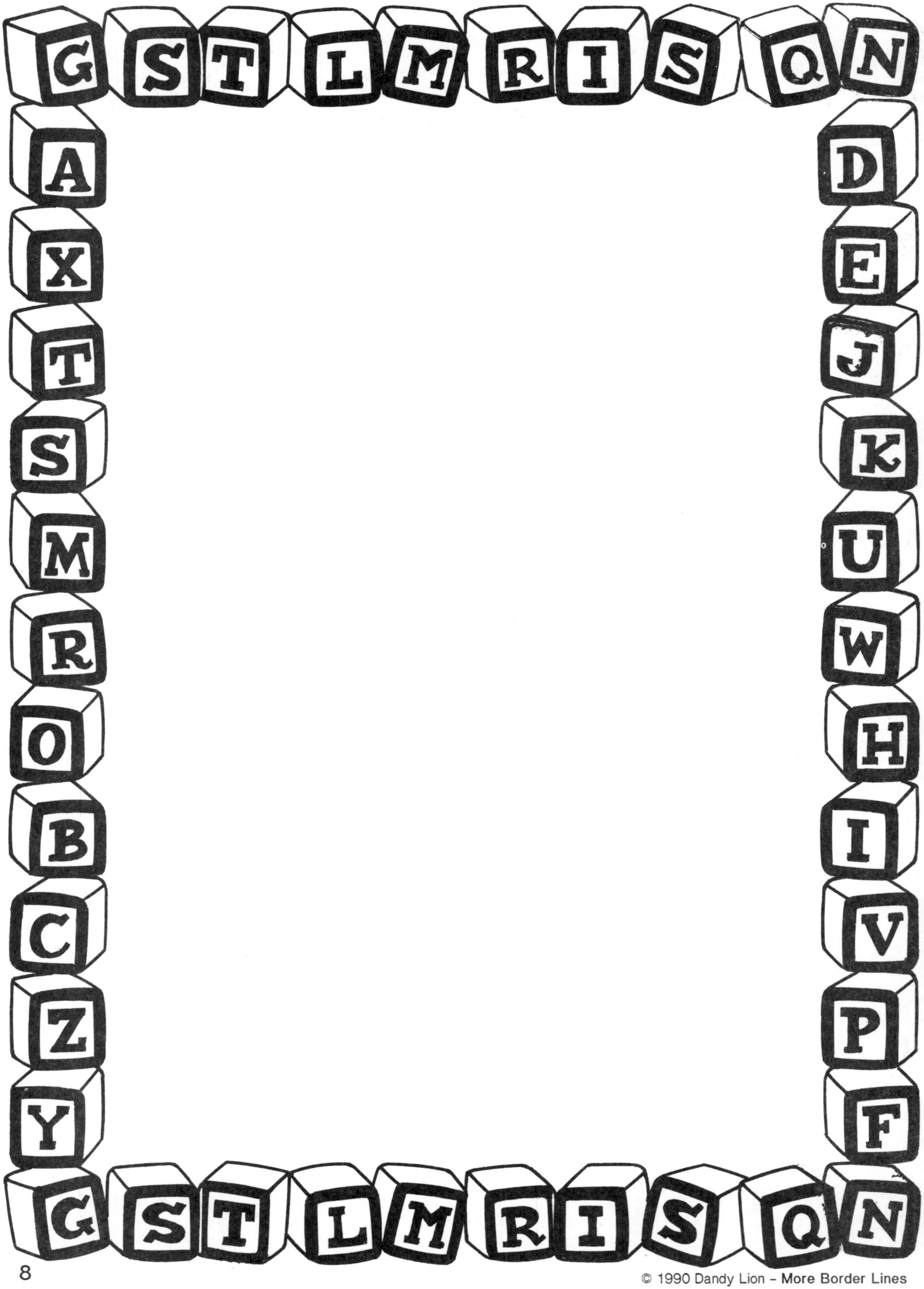
G S T L M R I S Q N
A X T S M R O B C Z Y
D E J K U W H I V P F
G S T L M R I S Q N

TALL TALES
WILD TALES
SAD TALES
RAFFLES
QUIZ ?
THE VIEW from HERE
DOG DAYS

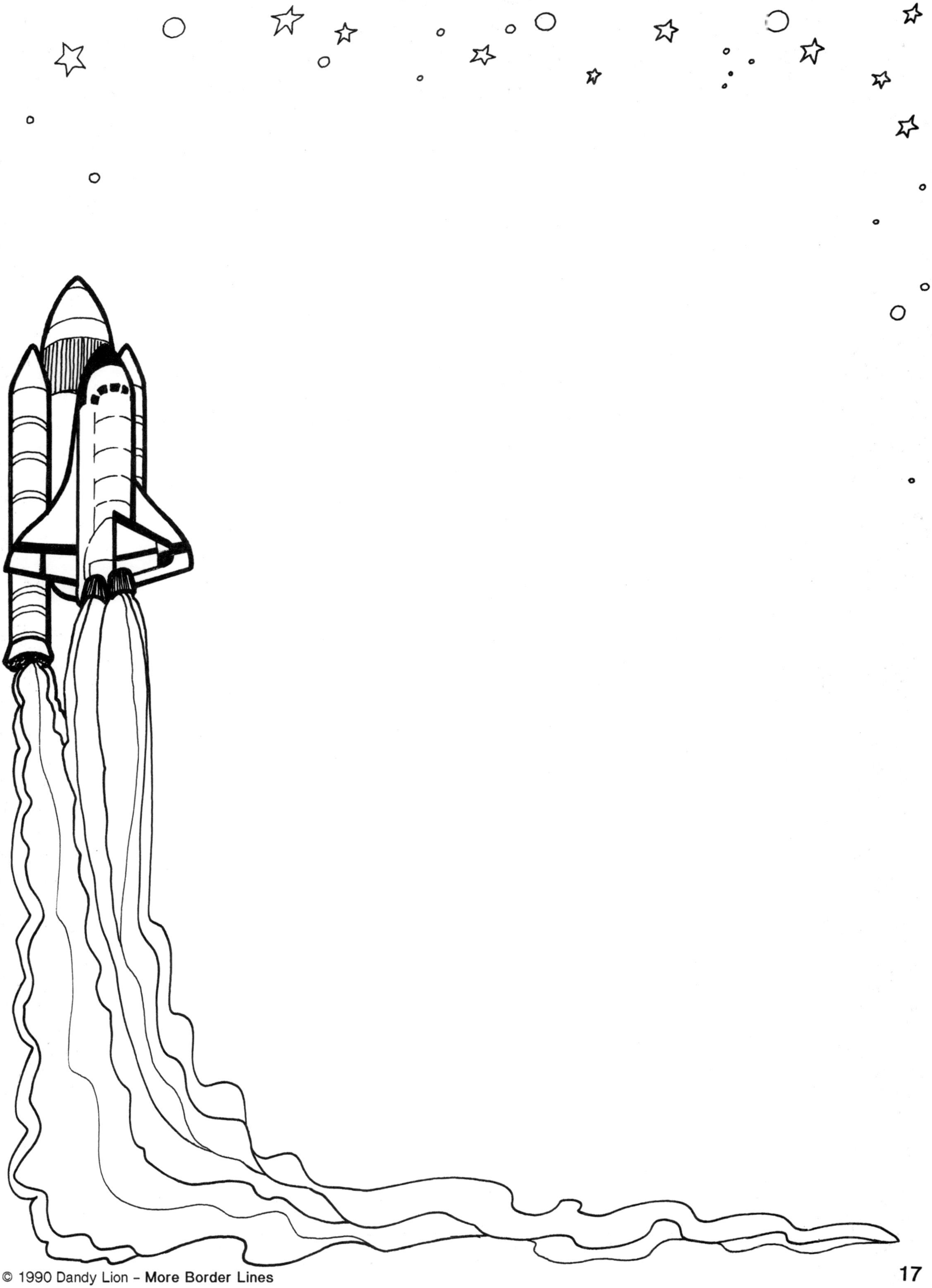

Sketch Book
Modeling Clay
Sketch Book
Modeling Clay

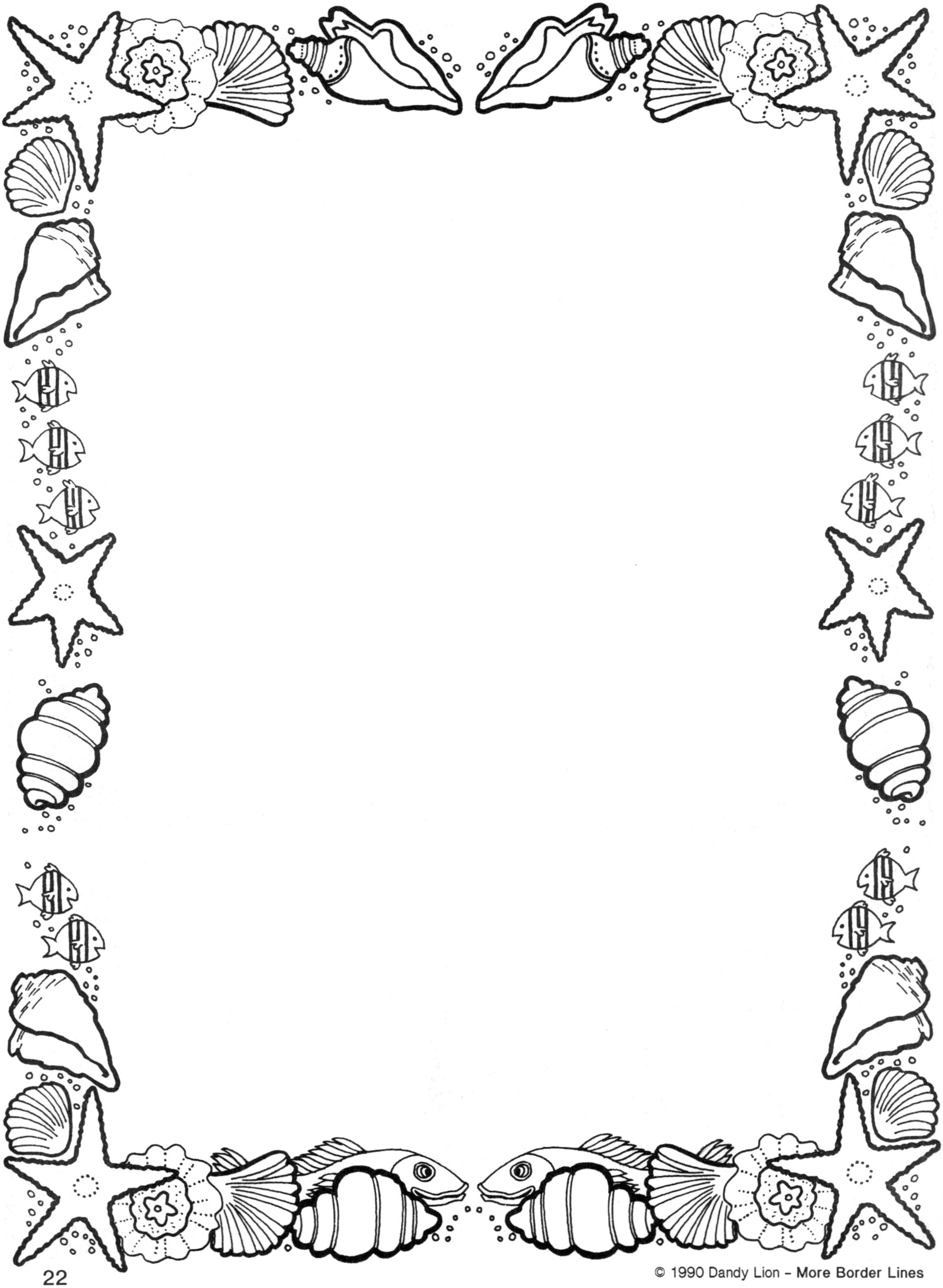

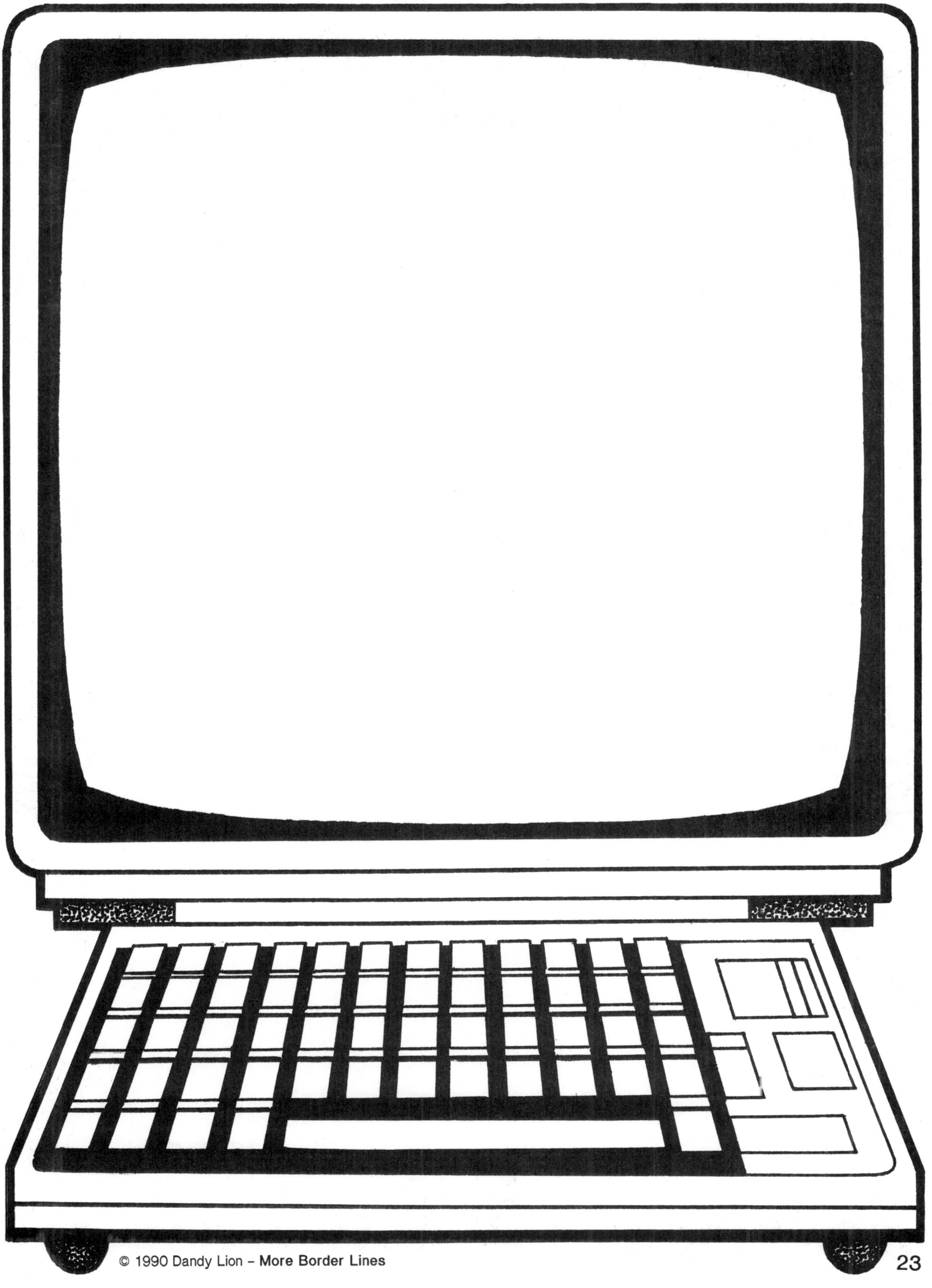

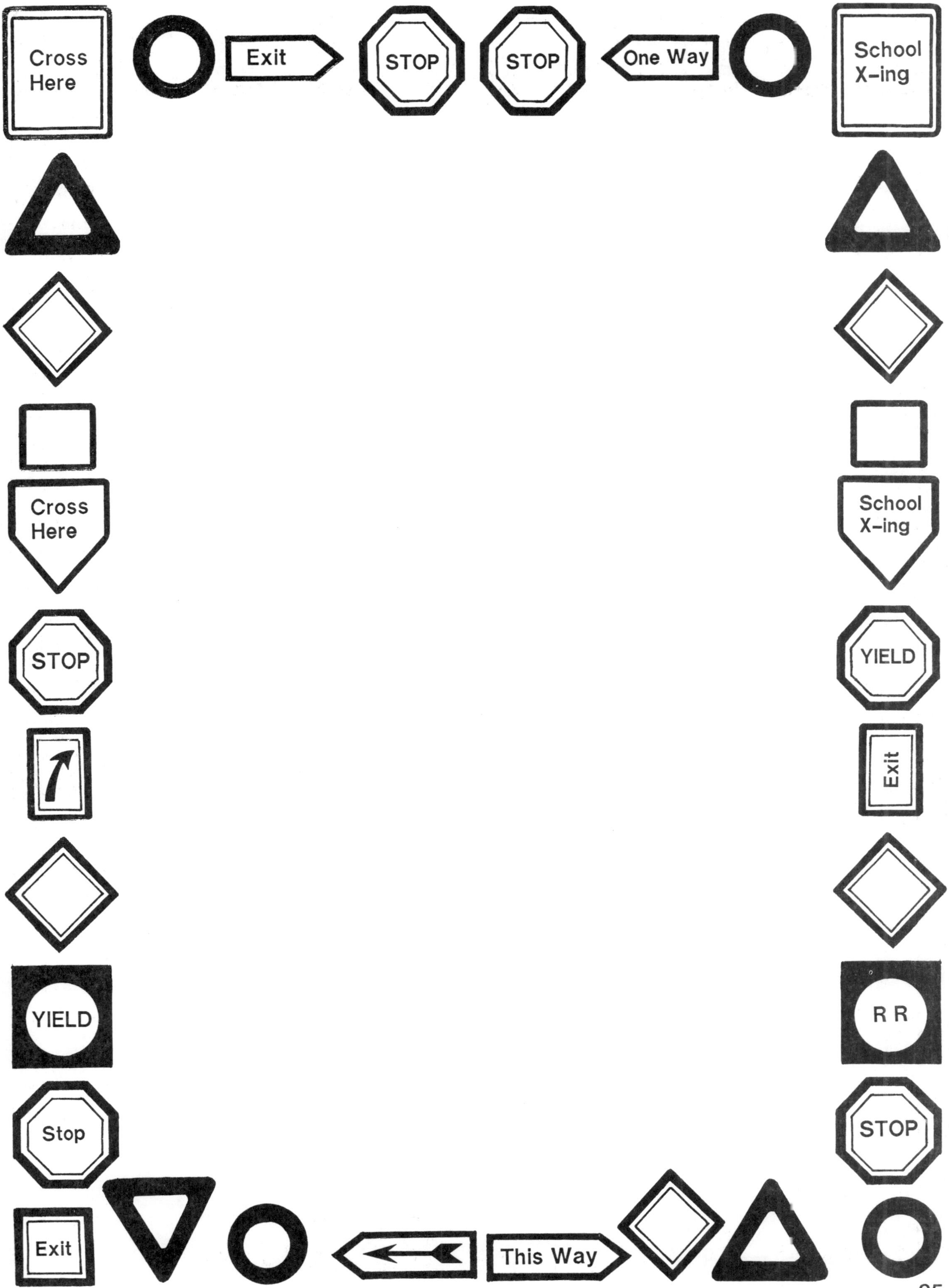
Cross Here
Exit
STOP
STOP
One Way
School X-ing
Cross Here
School X-ing
STOP
YIELD
Exit
YIELD
R R
Stop
STOP
Exit
This Way

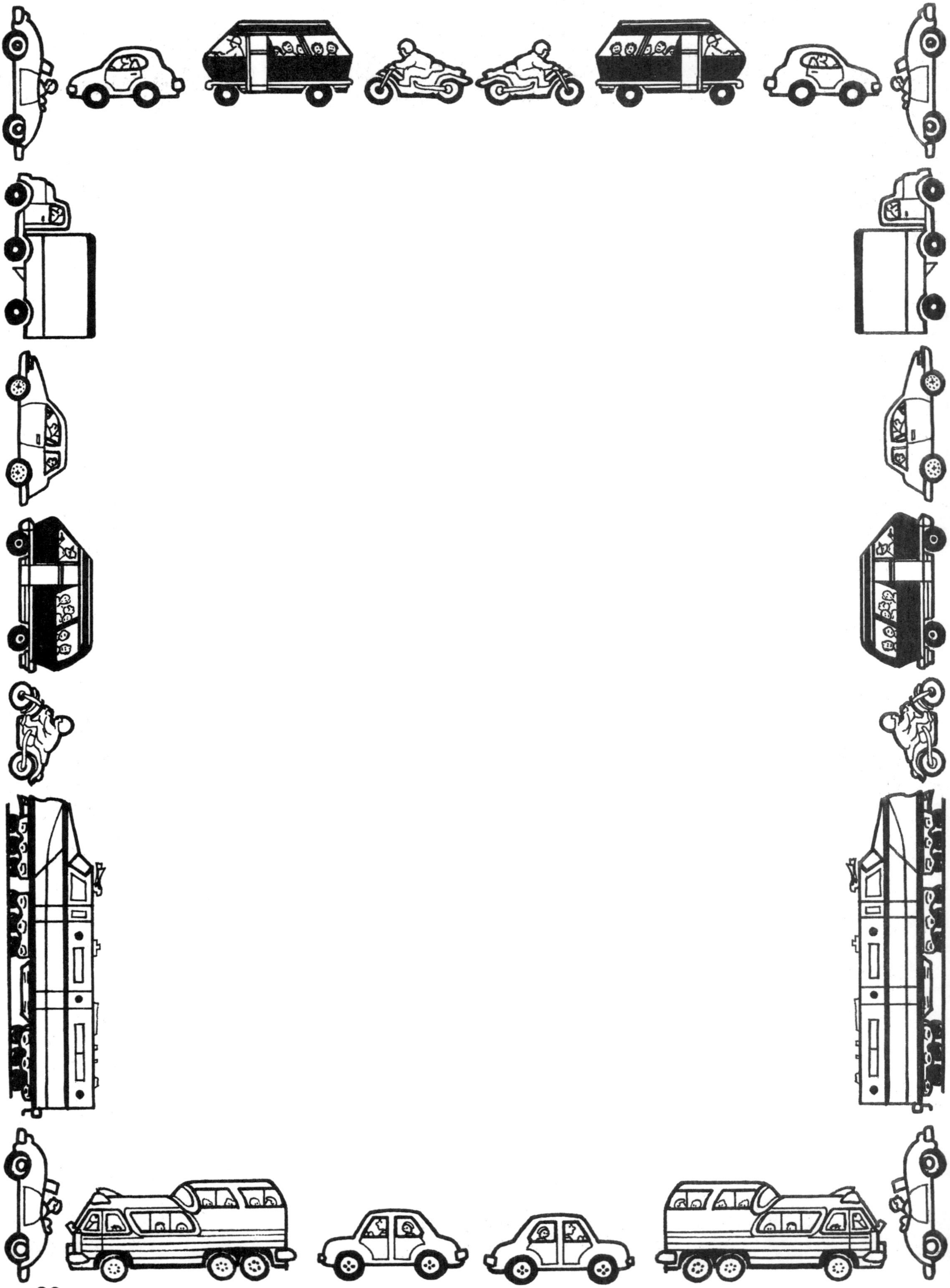

AbCDefGHIJKLMN
OPQrSTUvWXYZ

The Most Intense Days

THE MOST INTENSE DAYS

Hand Calculate Your Primary Directions

John Savarese

Published by: John Savarese
August 2016
Author: John Savarese.
Front Cover drawing: John Savarese; planet symbols courtesy of Solar Fire.

ISBN-13: 978-0-692-76161-8

Website: www.myprimaries.com

The astrological charts in this book were generated using Solar Fire astrology software (www.astrolabe.com).

The astrological speculums in this book were generated using Janus astrology software (www.astrology-house.com) v4.3, Astrology House, 147A Centreway Road, Orewa 0931, Auckland, New Zealand.

Contents

Preface

In a person's lifetime, each planetary body in our solar system will have its particular day, that is, when the effect of that planetary body will be most pronounced and maximized. For example, it is useful to know the date when Jupiter will be most powerful in your life to take advantage of all the benefits it bestows. Had I known this date, I would have undoubtedly bought some risky real estate offered to me at that time which turned out to be quite valuable a short time later. I have no regrets however. I do know the future dates when other primary directions will occur. After you read this book and implement its techniques, you too will know when beneficial primary directions will occur in your life so you can take advantage of them. You will also know when some not so beneficial primary directions will occur. Then you will be able to plan ahead and deal with their challenges.

To hand-calculate primary directions, you must first know how the angles in a chart are directed to the planets in the chart resulting in 'directions' which when applied to a person's life become primary directions. If you are not completely familiar with directing angles to planets, I highly recommend that you read *Western Sidereal Astrology for Beginners* by John Savarese available from www.myprimaries.com and www.amazon.com which completely describes how angles are directed to planets in easy-to-understand clear language. To get a more detailed approach, see the trailblazing book *An Introduction to Western Sidereal Astrology* by Kenneth Bowser available on www.amazon.com.

—John Savarese, B.A. Syracuse University 1972, M.A. City University of New York, Hunter College 1974, NCGR-PAA (National Council for Geocosmic Research-Professional Astrologers' Alliance) Level 3.

August 2016, New York City

Chapter 1

THE TECHNIQUE

A primary directed planet will tend to be manifesting greatly in your life at the age when it occurs. You must prepare in advance for the occurrence of a primary direction to be able to reap its benefits or to avoid its pitfalls. You might want to know when Saturn will tend to manifest greatly in your life so you can prepare for the challenges it will bring. You might also want to know when Jupiter will tend to manifest greatly in your life so you can take advantage of the benefits it bestows. Here are some examples of how primary directed planets will influence your life:

- *Sun* Your brilliance, willpower, creativity and vitality shine in your new world.
- *Moon* Family, home, and health matters come to the fore—let your instincts (as well as your rationality) guide you to success.
- *Mercury* Use your mind (communication and cognition) and get moving to launch that business, book, college degree or project.
- *Venus* A very pleasurable and peaceful time in your life (and appreciate the beauty of the experience).
- *Mars* Ride the energy wave, stay local, deal with possible inflammation and fever issues, and be cool (low profile).
- *Jupiter* Good luck and great times (but don't go overboard).
- *Saturn* Not such great times but if you get serious, work hard and persevere to get through the problems and delays, you will be amply rewarded.

- *Uranus* Excitement, surprises and spontaneous changes in your new, unique, independent world.
- *Neptune* Experience mysticism and spirituality, music and the arts, or drift into illusion and escapism.
- *Pluto* A major crisis; deal with it as well as you can to get through it.

For many more meanings of these planetary bodies and their effects during primary directions, consult *The Astrology Sourcebook: Your Guide to Understanding the Symbolic Language of the Stars* by Shirley Soffer available at www.amazon.com.

How well a planet manifests depends on its natal dignity or debility (this does not apply to the primary directions of stars as they have their own fixed meanings without having dignities or debilities). When the primary direction of a planet occurs, other conditions at that time such as the transits at that time, transits to the natal planets at that time, the solar return in force at that time, the lunar return in force at that time, progressed planets at that time, elevations of the planets in the chart at that time, and any other influences you may prefer to include, will also come into play. For a primary directed star, the meaning of that star will determine how it manifests.

Using a Casio fx-260 Solar calculator, you can calculate the amount of time it takes for an angle to be directed to a planet by noting the sidereal time of the planet from the column in the speculum of the angle being directed (this will be shown in the Person E example below and the examples that follow Person E). Then from that sidereal time shown in hours:minutes:seconds, subtract the natal LST in hours:minutes:seconds displayed at the top of the speculum. This will give you the amount of time during the sidereal day of your birth that it will take for the direction to occur to that planet. This amount of time is known as the arc of direction. Change the hours:minutes:seconds of the arc of direction to minutes:seconds. Then change these minutes:seconds to a decimal and divide the decimal amount by four.

The result is the actual age (for example, age 7 or 53.5 (age 53 and one half) at which the angle will be directed to that planet in your life. The properties of the planet will have a tendency to manifest greatly in your life at that age. The arc of direction is the total amount of time needed for the angle to be directed to the planet you selected, and this arc of direction amount of time must be divided by four be-

cause every four minute increment in this amount of time starting from your LST is equal to one year of your life. The number of four-minute increments starting from the LST is your age at which the angle will be directed to the planet. This type of direction that occurs during a person's lifetime is known as a primary direction.

To get the first primary direction to a planet that will occur in your life in order to calculate the actual age and date on which it will occur, look at your birthchart and choose the planet that, going clockwise, will hit an angle first (i.e. when that angle will be directed to it). To get the second primary direction to a planet that will occur in your life in order to calculate the actual age and date on which it will occur, look at your birthchart and choose the planet that, going clockwise, will hit an angle next (i.e. when that angle will be directed to it). Do this in sequence for all the planets in your chart.

Note that in the examples which follow, if the LST is greater than the sidereal time value of the planet you are subtracting the LST from, just add 24:00:00 to the sidereal time value and then subtract the LST. Also, if you get an outlandish age for a primary direction, the planet you see in the chart is actually already past the angle you wanted to direct the planet to, even though it looks like the planet is coming to the angle in the chart. In this case, direct the next angle in the chart to the planet, for example, use the IC instead of the Descendant.

Chapter 2

THE PRIMARY DIRECTION OF THE DESCENDANT TO MERCURY FOR PERSON E

This primary direction data is taken from Chapter 6 – Primary Directions in Your Life from *Western Sidereal Astrology for Beginners* by John Savarese. In that chapter, you were shown how to hand-calculate the age at which the Descendant is directed to Mercury. Now in this chapter and in the following chapters, you will be shown not only how to hand-calculate the ages at which primary directions occur but also the actual dates on which they occur.

First, create the chart of Person E, whose birthdata is August 4, 1959, 2:30 pm, Brooklyn, NY. We are using Solar Fire to create this chart, which is shown on the next page.

Person E. Natal Chart. Aug. 4, 1959. 2:30 PM, EDT +04:00:00, Brooklyn, NY, USA,Geocentric, Western Sidereal (Fagan-Allen), Campanus, True Node.

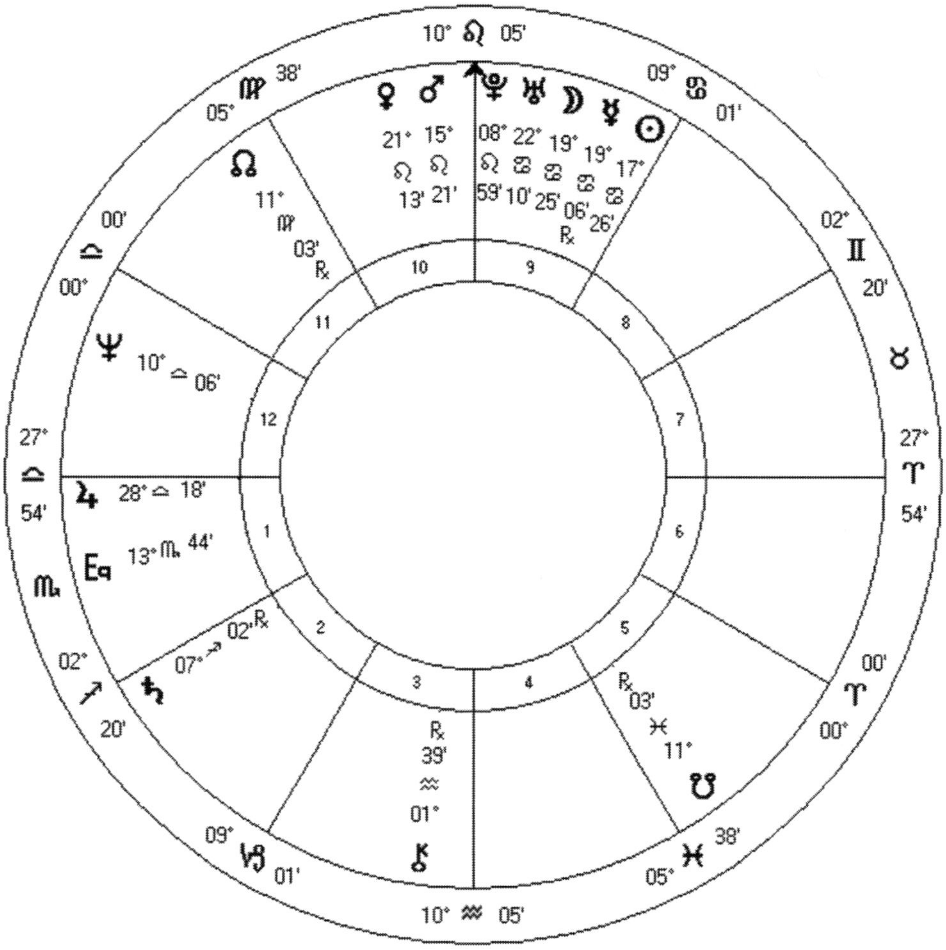

Then cast the chart for the birthdata of Person E using Janus. After you cast the chart and it is displayed, you must create the astrological speculum of Person E. Go to the Calculate dropdown and select Diurnal Events (Parans), then from the Settings dropdown, select Options, then select Planet on Angle and hit OK. The following speculum displays.

Person E. Speculum. Aug 4, 1959. 2:30 PM, EST +04:00:00, Brooklyn, NY, USA, Geocentric, Western Sidereal (Fagan-Bradley) Campanus, True Node.

RAMC 156 08' 31" LST 10:24:34

Diurnal Events: Planet on Angle
Sidereal Time

Planet/Star	Rise	Mc	Set	Ic
Sun	1:54:19	8:56:20	15:58:21	20:56:20
Moon	2:14:10	8:59:33	15:44:55	20:59:33
Mercury	2:14:49	8:57:17	15:39:45	20:57:17
Venus	4:52:58	11:00:06	17:07:15	23:00:06
Mars	4:15:00	10:45:55	17:16:50	22:45:55
Jupiter	10:24:20	15:21:11	20:18:02	3:21:11
Saturn	13:28:56	18:05:17	22:41:37	6:05:17
Uranus	2:16:57	9:16:05	16:15:12	21:16:05
Neptune	8:49:59	14:10:31	19:31:03	2:10:31
Pluto	3:20:27	10:38:35	17:56:43	22:38:35
Moon's North Node	6:26:22	12:19:13	18:12:04	0:19:13
Aldebaran	21:37:28	4:36:28	11:35:29	16:36:28

Here is the Casio FX-260 Solar calculator:

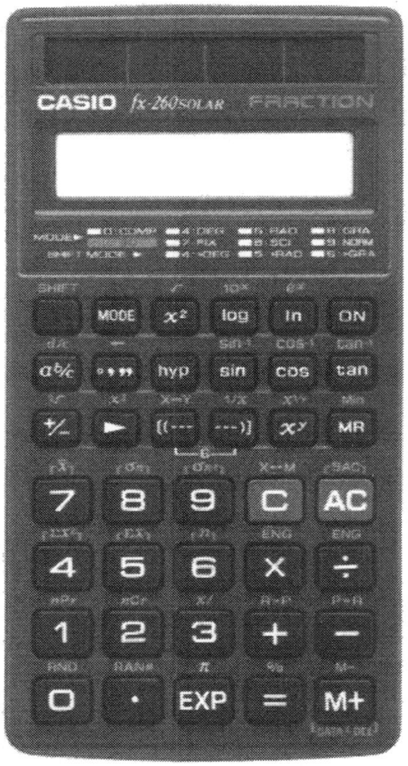

Before we begin, you must know how to convert degrees, minutes and seconds to decimal form and then back again to degrees, minutes and seconds.

Type 15 for degrees, hit the degrees button [° ’ ”], type 39 for minutes, hit the degrees button [° ’ ”], type 45 for seconds. Hit the Shift button [SHIFT] located in the upper left corner of the calculator and then hit the degrees button [° ’ ”] to get the decimal form of the degrees, minutes and seconds value. When a decimal value is shown, hit the Shift button [SHIFT] and then the degrees button [° ’ ”] to go back to degrees, minutes and seconds.

Now to begin.

First, we convert the date and the year of Person E's birth to decimal form.

For August 4th, there are 27 days until the end of August (to August 31st).

To 27, add the 30 days of September, the 31 days of October, the 30 days of November and the 31 days of December. The total is 149 days left in the year from August 4 to and including December 31.

Divide 149 by 365 (because 1959 is not a leap year; if it was a leap year you would divide by 366):

149/365 equals .4

Make .4 negative by hitting the plus/minus key on the calculator, so you have -.4

Then add 1.000 to -.4 to get .6

We now have the birthday in decimal form: 1959.6

Calculate the number of years and days from the birth of Person E for the Descendant to Mercury primary direction to occur.

Take the sidereal time value from the speculum when the Descendant is directed to Mercury, by looking at the row of Mercury and then down the Descendant (Set) column. The sidereal time is 15h39m45s. Also note the LST of the birth you get at the top of the speculum, which is 10h24m34s.

Type 15 hit the degrees button, type 39 hit the degrees button, type 45

Hit the subtract button.

Type 10 hit the degrees button, type 24 hit the degrees button, type 34

Hit the equal button to get 5h15m11s.

By hand, change 5h15m11s to minutes and seconds. This equals 315m11s.

Change 315m11s to a decimal and divide by 4 by doing the following:

Type 315 hit the degrees button, type 11

Hit the Shift button located in the upper left corner of the calculator.

Then hit the degrees button to get 315.1833333 which is the decimal.

Hit the divide button, type 4 and hit the equal button.

You will get 78.79583333 which rounded off is 78.8.

This is the age at which Mercury will tend to manifest greatly in Person E's life. Mercury means business, among other things. Since Person E's Mercury is in Cancer in the chart, perhaps the primary direction indicates that Person E's family house will be sold (Cancer can signify the family, the home and property). If Mars comes into play at that age, there may be a family conflict in the sale. If Venus comes into play, the sale will probably go smoothly.

Now to get the date on which this occurs, we add the birthday in decimal form to the date of the primary direction in decimal form:

1959.6 plus 78.8 equals 2038.4

This is the date on which the primary direction occurs in decimal form.

We now convert the decimal form to the actual date.

x/365 = .4/1

Solving for x, the result is 146.

Look at the following table:

Month	*Number of Days in the Month*	*Subtotal of Days*	*Subtotal of Days for a Leap Year*
Jan	31	31	31
Feb	28	59	60
Mar	31	90	91
April	30	120	121
May	31	151	152
June	30	181	182
July	31	212	213
Aug	31	243	244
Sep	30	273	274
Oct	31	304	305
Nov	30	334	335
Dec	31	365	366

X = 146 so 146 minus 120 for April equals 26 for May. Thus, the actual date on which the primary direction occurs is May 26, 2038.

Chapter 3

LEAP YEARS

You will need to know when leap years occur:

1900 (leap year)
1901
1902
1903
1904 (leap year)
1905
1906
1907
1908 (leap year)
1909
1910
1911
1912 (leap year)
1913
1914
1915
1916 (leap year)
1917
1918
1919
1920 (leap year)
1921
1922
1923
1924 (leap year)
1925
1926
1927
1928 (leap year)
1929
1930
1931
1932 (leap year)
1933
1934
1935
1936 (leap year)
1937
1938
1939
1940 (leap year)
1941
1942
1943
1944 (leap year)
1945
1946
1947
1948 (leap year)
1949
1950
1951
1952 (leap year)
1953
1954
1955
1956 (leap year)

1957
1958
1959
1960 (leap year)
1961
1962
1963
1964 (leap year)
1965
1966
1967
1968 (leap year)
1969
1970
1971
1972 (leap year)
1973
1974
1975
1976 (leap year)
1977
1978
1979
1980 (leap year)
1981
1982
1983
1984 (leap year)
1985
1986
1987
1988 (leap year)
1989
1990
1991
1992 (leap year)
1993
1994
1995
1996 (leap year)
1997
1998
1999
2000 (leap year)
2001
2002
2003
2004 (leap year)
2005
2006
2007
2008 (leap year)
2009
2010
2011
2012 (leap year)
2013
2014
2015
2016 (leap year)
2017
2018
2019
2020 (leap year)
2021
2022
2023
2024 (leap year)
2025
2026
2027
2028 (leap year)
2029
2030
2031
2032 (leap year)
2033
2034
2035
2036 (leap year)
2037
2038
2039
2040 (leap year)
2041
2042
2043
2044 (leap year)
2045
2046
2047
2048 (leap year)
2049
2050
2051
2052 (leap year)
and so on.

Chapter 4

THE PRIMARY DIRECTIONS OF PERSON AA

First, create the chart of Person AA, whose birthdata is March 19, 1971, 11:30 am, Chicago IL. We are using Solar Fire to create this chart, which is shown on the next page.

Person AA. Natal Chart. Mar 19, 1971.11:30 AM, CST +06:00:00, Chicago IL, USA,Geocentric, Western Sidereal (Fagan-Allen), Campanus, True Node.

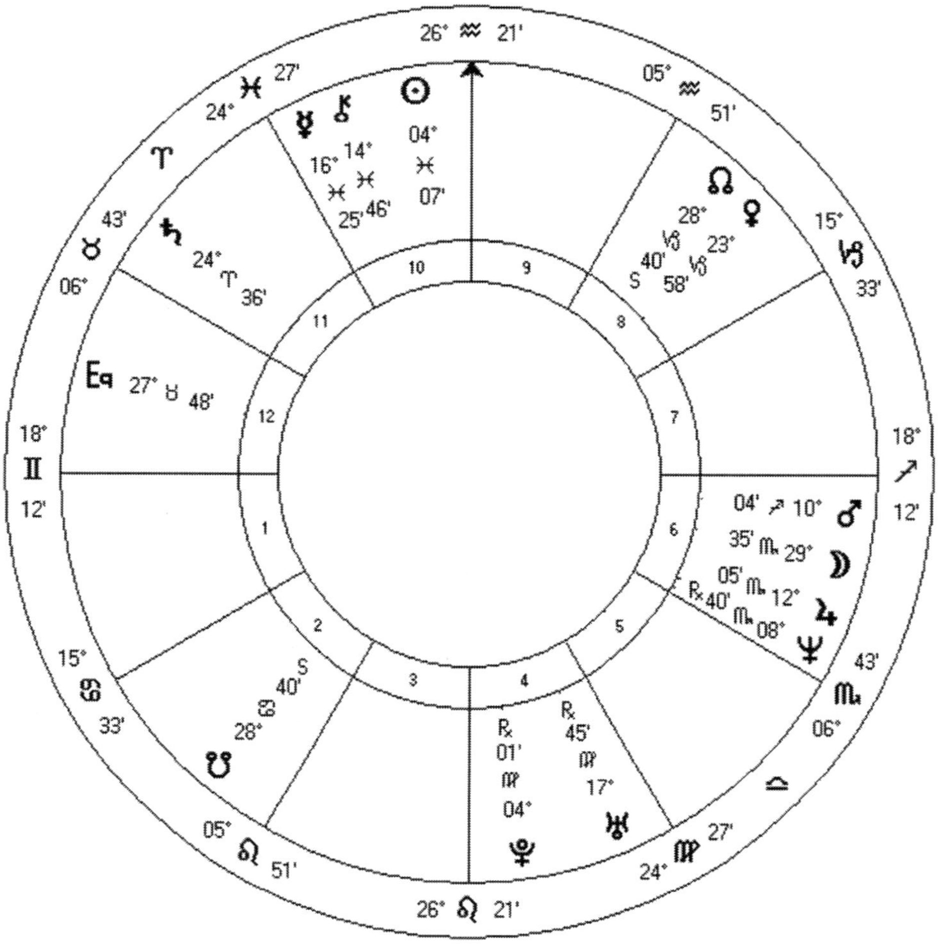

Then cast the chart for the birthdata of Person AA using Janus. After you cast the chart and it is displayed, you must create the astrological speculum of Person AA. Go to the Calculate dropdown and select Diurnal Events (Parans), then from the Settings dropdown, select Options, then select Planet on Angle and hit OK. The following speculum displays.

Person AA. Speculum. Mar 19, 1971. 11:30 AM, CST +06:00:00, Chicago IL, USA, Geocentric, Western Sidereal (Fagan-Bradley) Campanus, True Node.

RAMC 345 32' 43" LST 23:02:10

Diurnal Events: Planet on Angle

Sidereal Time

Planet/Star	Rise	Mc	Set	Ic
Sun	17:43:06	23:32:25	5:21:43	11:32:25
Moon	6:58:03	12:31:02	18:04:01	0:31:02
Mercury	18:02:11	23:58:00	5:53:48	11:58:00
Venus	15:58:03	20:54:22	1:50:41	8:54:22
Mars	13:35:12	18:03:27	22:31:42	6:03:27
Jupiter	11:36:20	16:18:28	21:00:36	4:18:28
Saturn	20:09:06	3:06:01	10:02:56	15:06:01
Uranus	7:01:39	12:46:32	18:31:24	0:46:32
Neptune	11:17:46	16:05:25	20:53:04	4:05:25
Pluto	5:23:17	12:21:35	19:19:54	0:21:35
Moon's North Node	16:33:31	21:43:01	2:52:30	9:43:01
Aldebaran	21:35:24	4:37:08	11:38:53	16:37:08

Now to begin.

First, we convert the date and the year of Person AA's birth to decimal form.

For March 19th, there are 12 days until the end of March (to and including March 31).

To 12, add the 30 days of April, 31 days of May, 30 days of June, 31 days of July, 31 days of August, 30 days of September, the 31 days of October, the 30 days of November and the 31 days of December. The total is 287 days left in the year from March 19 up to and including December 31.

Divide 287 by 365 (because 1971 is not a leap year; if it was a leap year you would divide by 366):

287/365 equals .786

Make .786 negative by hitting the +/- plus/minus key on the calculator, so you have -.786

Then add 1.000 to -.786 to get .214

We now have the birthday in decimal form:

1971.214

1. Calculate the number of years and days from the birth of Person AA for the Midheaven to the Sun primary direction to occur.

Take the sidereal time value from the speculum when the Midheaven is directed to the Sun, by looking at the row of the Sun and then down the Midheaven (Mc) column. The sidereal time is 23h32m35s. Also note the LST of the birth you get at the top of the speculum, which is 23h02m10s.

Type 23 hit the degrees button ° ' ", type 32 hit the degrees button ° ' ", type 35

Hit the − subtract button.

Type 23 hit the degrees button ° ' ", type 2 hit the degrees button ° ' ", type 10

Hit the equal button = to get 0h30m25s.

By hand, change 0h30m25s to minutes and seconds. This equals 30m25s.

Change 30m25s to a decimal and divide by 4 by doing the following:

Type 30 hit the degrees button ° ' ", type 25

Hit the Shift button located in the upper left corner of the calculator.

Then hit the degrees button to get 30.41666 which is the decimal.

Hit the divide button , type 4 and hit the equal button .

You will get 7.6.

This is the age at which the Sun will tend to manifest greatly in Person AA's life.

Now to get the date on which this occurs, we add the birthday in decimal form to the date of the primary direction in decimal form:

1971.214 plus 7.6 equals 1978.814

This is the date on which the primary direction occurs in decimal form.

We now convert the decimal form to the actual date.

x/365 = .814/1

Solving for x, the result is 297.

Look at the following table:

Month	*Number of Days in the Month*	*Subtotal of Days*	*Subtotal of Days for a Leap Year*
Jan	31	31	31
Feb	28	59	60
Mar	31	90	91
April	30	120	121
May	31	151	152
June	30	181	182
July	31	212	213
Aug	31	243	244
Sep	30	273	274
Oct	31	304	305
Nov	30	334	335
Dec	31	365	366

As x = 297 so 297 minus 273 for Sep equals 24 for Oct, thus the actual date on which the primary direction occurs is Oct. 24, 1978.

2. Calculate the number of years and days from the birth of Person AA for the Midheaven to Mercury primary direction to occur.

Take the sidereal time value from the speculum when the Midheaven is directed to Mercury, by looking at the row of Mercury and then down the Midheaven (Mc) column. The sidereal time is 23h58m0s. Also note the LST of the birth you get at the top of the speculum, which is 23h02m10s.

Type 23 hit the degrees button, type 59 hit the degrees button, type 0

Hit the subtract button.

Type 23 hit the degrees button, type 2 hit the degrees button, type 10

Hit the equal button to get 0h55m50s.

By hand, change 0h55m50s to minutes and seconds. This equals 55m50s.

Change 55m50s to a decimal and divide by 4 by doing the following:

Type 55 hit the degrees button, type 50

Hit the Shift button located in the upper left corner of the calculator.

Then hit the degrees button to get 55.8333 which is the decimal.

Hit the divide button, type 4 and hit the equal button.

You will get 14.

This is the age at which Mercury will tend to manifest greatly in Person AA's life.

Now to get the date on which this occurs, we add the birthday in decimal form to the date of the primary direction in decimal form:

1971.214 plus 14 equals 1985.214

This is the date on which the primary direction occurs in decimal form.

We now convert the decimal form to the actual date.

x/365 = .214/1

Solving for x, the result is 78.

Look at the following table:

Month	*Number of Days in the Month*	*Subtotal of Days*	*Subtotal of Days for a Leap Year*
Jan	31	31	31
Feb	28	59	60
Mar	31	90	91
April	30	120	121
May	31	151	152
June	30	181	182
July	31	212	213
Aug	31	243	244
Sep	30	273	274
Oct	31	304	305
Nov	30	334	335
Dec	31	365	366

As x = 78 so 78 minus 59 for Feb equals 19 for March, thus the actual date on which the primary direction occurs is March 19, 1985.

3. Calculate the number of years and days from the birth of Person AA for the IC to Pluto primary direction to occur.

Take the sidereal time value from the speculum when the IC is directed to Pluto, by looking at the row of Pluto and then down the IC column. The sidereal time is 0h21m35s. Also note the LST of the birth you get at the top of the speculum, which is 23h02m10s.

Because the LST is greater than the sidereal time, add 24h0m0s to 0hr21m35s to get 24h21m35s.

Type 24 hit the degrees button [° ′ ″], type 21 hit the degrees button [° ′ ″], type 35

Hit the [−] subtract button.

Type 23 hit the degrees button [° ′ ″], type 2 hit the degrees button [° ′ ″], type 10

Hit the equal button  to get 1h19m25s.
By hand, change 1h19m25s to minutes and seconds. This equals 79m25s.
Change 79m25s to a decimal and divide by 4 by doing the following:
Type 79 hit the degrees button , type 25
Hit the Shift button  located in the upper left corner of the calculator.
Then hit the degrees button  to get 79.41666 which is the decimal.
Hit the divide button 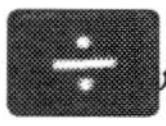, type 4 and hit the equal button .
You will get 19.854.
This is the age at which Pluto will tend to manifest greatly in Person AA's life.

Now to get the date on which this occurs, we add the birthday in decimal form to the date of the primary direction in decimal form:
1971.214 plus 19.854 equals 1991.068
This is the date on which the primary direction occurs in decimal form.
We now convert the decimal form to the actual date.
x/365 = .068/1
Solving for x, the result is 25.
Look at the following table:

Month	*Number of Days in the Month*	*Subtotal of Days*	*Subtotal of Days for a Leap Year*
Jan	31	31	31
Feb	28	59	60
Mar	31	90	91
April	30	120	121
May	31	151	152
June	30	181	182
July	31	212	213
Aug	31	243	244
Sep	30	273	274
Oct	31	304	305
Nov	30	334	335
Dec	31	365	366

As x = 25 so 25 minus zero is 25, thus the actual date on which the primary direction occurs is Jan 25, 1991.

4. Calculate the number of years and days from the birth of Person AA for the IC to Moon primary direction to occur.

Take the sidereal time value from the speculum when the IC is directed to the Moon, by looking at the row of the Moon and then down the IC column. The sidereal time is 0h31m02s. Also note the LST of the birth you get at the top of the speculum, which is 23h02m10s.

Because the LST is greater than the sidereal time, add 24h0m0s to 0hr31m02s to get 24h31m02s.

Type 24 hit the degrees button , type 31 hit the degrees button , type 2

Hit the subtract button.

Type 23 hit the degrees button , type 2 hit the degrees button , type 10

Hit the equal button to get 1h28m52s.

By hand, change 1h28m52s to minutes and seconds. This equals 88m52s.

Change 88m52s to a decimal and divide by 4 by doing the following:

Type 88 hit the degrees button , type 52

Hit the Shift button located in the upper left corner of the calculator.

Then hit the degrees button to get 88.866 which is the decimal.

Hit the divide button , type 4 and hit the equal button .

You will get 22.216.

This is the age at which the Moon will tend to manifest greatly in Person AA's life.

Now to get the date on which this occurs, we add the birthday in decimal form to the date of the primary direction in decimal form:

1971.214 plus 22.216 equals 1993.43

This is the date on which the primary direction occurs in decimal form.

We now convert the decimal form to the actual date.

x/365 = .43/1

Solving for x, the result is 157.

Look at the following table:

Month	*Number of Days in the Month*	*Subtotal of Days*	*Subtotal of Days for a Leap Year*
Jan	31	31	31
Feb	28	59	60
Mar	31	90	91
April	30	120	121
May	31	151	152
June	30	181	182
July	31	212	213
Aug	31	243	244
Sep	30	273	274
Oct	31	304	305
Nov	30	334	335
Dec	31	365	366

As x = 157 so 157 minus 151 for May equals 6 for June, thus the actual date on which the primary direction occurs is June 6, 1993.

5. Calculate the number of years and days from the birth of Person AA for the IC to Uranus primary direction to occur.

Take the sidereal time value from the speculum when the IC is directed to Uranus, by looking at the row of Uranus and then down the IC column. The sidereal time is 0h46m32s. Also note the LST of the birth you get at the top of the speculum, which is 23h02m10s.

Because the LST is greater than the sidereal time, add 24h0m0s to 0hr46m32s to get 24h46m32s.

Type 24 hit the degrees button ° ' '', type 46 hit the degrees button ° ' '', type 32

Hit the − subtract button.

Type 23 hit the degrees button ° ' '', type 2 hit the degrees button ° ' '', type 10

Hit the equal button  to get 1h44m22s.
By hand, change 1h44m22s to minutes and seconds. This equals 104m22s.
Change 104m22s to a decimal and divide by 4 by doing the following:
Type 104 hit the degrees button , type 22
Hit the Shift button  located in the upper left corner of the calculator.
Then hit the degrees button  to get 104.3666 which is the decimal.
Hit the divide button , type 4 and hit the equal button .
You will get 26.
This is the age at which Uranus will tend to manifest greatly in Person AA's life.

Now to get the date on which this occurs, we add the birthday in decimal form to the date of the primary direction in decimal form:
1971.214 plus 26 equals 1997.214
This is the date on which the primary direction occurs in decimal form.
We now convert the decimal form to the actual date.
x/365 = .214/1
Solving for x, the result is 78.
Look at the following table:

Month	*Number of Days in the Month*	*Subtotal of Days*	*Subtotal of Days for a Leap Year*
Jan	31	31	31
Feb	28	59	60
Mar	31	90	91
April	30	120	121
May	31	151	152
June	30	181	182
July	31	212	213
Aug	31	243	244
Sep	30	273	274
Oct	31	304	305
Nov	30	334	335
Dec	31	365	366

As x = 78 so 78 minus 59 for Feb equals 19 for March, thus the actual date on which the primary direction occurs is March 19, 1997.

6. Calculate the number of years and days from the birth of Person AA for the Descendant to Venus primary direction to occur.

Take the sidereal time value from the speculum when the Descendant is directed to Venus, by looking at the row of Venus and then down the Descendant (Set) column. The sidereal time is 1h50m41s. Also note the LST of the birth you get at the top of the speculum, which is 23h02m10s.

Because the LST is greater than the sidereal time, add 24h0m0s to 1hr50m41s to get 25h50m44s.

Type 25 hit the degrees button, type 50 hit the degrees button, type 44

Hit the subtract button.

Type 23 hit the degrees button, type 2 hit the degrees button, type 10

Hit the equal button to get 2h48m31s.

By hand, change 2h48m31s to minutes and seconds. This equals 168m31s.

Change 168m31s to a decimal and divide by 4 by doing the following:

Type 168 hit the degrees button, type 31

Hit the Shift button located in the upper left corner of the calculator.

Then hit the degrees button to get 168.516 which is the decimal.

Hit the divide button, type 4 and hit the equal button.

You will get 42.129.

This is the age at which Venus will tend to manifest greatly in Person AA's life.

Now to get the date on which this occurs, we add the birthday in decimal form to the date of the primary direction in decimal form:

1971.214 plus 42.129 equals 2013.34

This is the date on which the primary direction occurs in decimal form.

We now convert the decimal form to the actual date.

x/365 = .34/1

Solving for x, the result is 124.

Look at the following table:

Month	*Number of Days in the Month*	*Subtotal of Days*	*Subtotal of Days for a Leap Year*
Jan	31	31	31
Feb	28	59	60
Mar	31	90	91
April	30	120	121
May	31	151	152
June	30	181	182
July	31	212	213
Aug	31	243	244
Sep	30	273	274
Oct	31	304	305
Nov	30	334	335
Dec	31	365	366

As x = 124 so 124 minus 120 for April equals 4 for May, thus the actual date on which the primary direction occurs is May 4, 2013.

7. Calculate the number of years and days from the birth of Person AA for the Midheaven to Saturn primary direction to occur.

Take the sidereal time value from the speculum when the Descendant is directed to Saturn, by looking at the row of Saturn and then down the Midheaven (Mc) column. The sidereal time is 3h06m01s. Also note the LST of the birth you get at the top of the speculum, which is 23h02m10s.

Because the LST is greater than the sidereal time, add 24h0m0s to 3hr06m01s to get 27h06m01s.

Type 27 hit the degrees button [° ’ ”], type 6 hit the degrees button [° ’ ”], type 1

Hit the [—] subtract button.

Type 23 hit the degrees button [° ’ ”], type 2 hit the degrees button [° ’ ”], type 10.

Hit the equal button to get 4h3m51s.

By hand, change 4h3m51s to minutes and seconds. This equals 243m51s.

Change 243m51s to a decimal and divide by 4 by doing the following:

Type 243 hit the degrees button, type 51

Hit the Shift button located in the upper left corner of the calculator.

Then hit the degrees button to get 243.85 which is the decimal.

Hit the divide button, type 4 and hit the equal button.

You will get 61.

This is the age at which Saturn will tend to manifest greatly in Person AA's life.

Now to get the date on which this occurs, we add the birthday in decimal form to the date of the primary direction in decimal form:

1971.214 plus 61 equals 2032.214

This is the date on which the primary direction occurs in decimal form.

We now convert the decimal form to the actual date.

x/365 = .214/1

Solving for x, the result is 78.

Look at the following table:

Month	*Number of Days in the Month*	*Subtotal of Days*	*Subtotal of Days for a Leap Year*
Jan	31	31	31
Feb	28	59	60
Mar	31	90	91
April	30	120	121
May	31	151	152
June	30	181	182
July	31	212	213
Aug	31	243	244
Sep	30	273	274
Oct	31	304	305
Nov	30	334	335
Dec	31	365	366

As x = 78 minus 59 for Feb equals 19 for March thus the actual date on which the primary direction occurs is March 19, 2032.

8. Calculate the number of years and days from the birth of Person AA for the IC to Neptune primary direction to occur.

Take the sidereal time value from the speculum when the IC is directed to Neptune, by looking at the row of Neptune and then down the IC column. The sidereal time is 4h05m25s. Also note the LST of the birth you get at the top of the speculum, which is 23h02m10s.

Because the LST is greater than the sidereal time, add 24h0m0s to 4hr05m25s to get 28h05m25s.

Type 28 hit the degrees button, type 5 hit the degrees button, type 25

Hit the subtract button.

Type 23 hit the degrees button, type 2 hit the degrees button, type 10

Hit the equal button to get 5h3m15s.

By hand, change 5h3m15s to minutes and seconds. This equals 303m15s.

Change 303m15s to a decimal and divide by 4 by doing the following:

Type 303 hit the degrees button, type 15

Hit the Shift button located in the upper left corner of the calculator.

Then hit the degrees button to get 303.25 which is the decimal.

Hit the divide button, type 4 and hit the equal button.

You will get 75.8125.

This is the age at which Neptune will tend to manifest greatly in Person AA's life.

Now to get the date on which this occurs, we add the birthday in decimal form to the date of the primary direction in decimal form:

1971.214 plus 75.8125 equals 2047.0265

This is the date on which the primary direction occurs in decimal form.

We now convert the decimal form to the actual date.

x/365 = .0265/1

Solving for x, the result is 10.

Look at the following table:

Month	*Number of Days in the Month*	*Subtotal of Days*	*Subtotal of Days for a Leap Year*
Jan	31	31	31
Feb	28	59	60
Mar	31	90	91
April	30	120	121
May	31	151	152
June	30	181	182
July	31	212	213
Aug	31	243	244
Sep	30	273	274
Oct	31	304	305
Nov	30	334	335
Dec	31	365	366

As x = 10 so 10 minus zero is 10 thus the actual date on which the primary direction occurs is Jan 10, 1991.

9. Calculate the number of years and days from the birth of Person AA for the IC to Jupiter primary direction to occur.

Take the sidereal time value from the speculum when the IC is directed to Jupiter, by looking at the row of Jupiter and then down the IC column. The sidereal time is 4h18m28s. Also note the LST of the birth you get at the top of the speculum, which is 23h02m10s.

Because the LST is greater than the sidereal time, add 24h0m0s to 4hr18m28s to get 28h18m28s.

Type 28 hit the degrees button [° ' "], type 18 hit the degrees button [° ' "], type 28

Hit the [−] subtract button.

Type 23 hit the degrees button [° ' "], type 2 hit the degrees button [° ' "], type 10

Hit the equal button to get 5h16m18s.

By hand, change 5h16m18s to minutes and seconds. This equals 316m18s.

Change 316m18s to a decimal and divide by 4 by doing the following:

Type 316 hit the degrees button, type 18

Hit the Shift button located in the upper left corner of the calculator.

Then hit the degrees button to get 316.3 which is the decimal.

Hit the divide button, type 4 and hit the equal button.

You will get 79.

This is the age at which Jupiter will tend to manifest greatly in Person AA's life.

Now to get the date on which this occurs, we add the birthday in decimal form to the date of the primary direction in decimal form:

1971.214 plus 79 equals 2050.214

This is the date on which the primary direction occurs in decimal form.

We now convert the decimal form to the actual date.

x/365 = .214/1

Solving for x, the result is 78

Look at the following table:

Month	*Number of Days in the Month*	*Subtotal of Days*	*Subtotal of Days for a Leap Year*
Jan	31	31	31
Feb	28	59	60
Mar	31	90	91
April	30	120	121
May	31	151	152
June	30	181	182
July	31	212	213
Aug	31	243	244
Sep	30	273	274
Oct	31	304	305
Nov	30	334	335
Dec	31	365	366

As x = 78 minus 59 for Feb equals 19 for March, thus the actual date on which the primary direction occurs is March 19, 2050.

10. Calculate the number of years and days from the birth of Person AA for the IC to Mars primary direction to occur.

Take the sidereal time value from the speculum when the IC is directed to Mars, by looking at the row of Mars and then down the IC column. The sidereal time is 6h3m27s. Also note the LST of the birth you get at the top of the speculum, which is 23h02m10s.

Because the LST is greater than the sidereal time, add 24h0m0s to 6hr3m27s to get 30h3m27s.

Type 30 hit the degrees button, type 3 hit the degrees button, type 27

Hit the subtract button.

Type 23 hit the degrees button, type 2 hit the degrees button, type 10

Hit the equal button to get 7h1m19s.

By hand, change 7h1m19s to minutes and seconds. This equals 421m19s.

Change 421m19s to a decimal and divide by 4 by doing the following:

Type 421 hit the degrees button, type 19

Hit the Shift button located in the upper left corner of the calculator.

Then hit the degrees button to get 421.131666 which is the decimal.

Hit the divide button, type 4 and hit the equal button.

You will get 105.33.

This is the age at which Mars will tend to manifest greatly in Person AA's life.

Now to get the date on which this occurs, we add the birthday in decimal form to the date of the primary direction in decimal form:

1971.214 plus 42.129 equals 2013.34

This is the date on which the primary direction occurs in decimal form.

We now convert the decimal form to the actual date.

x/365 = .34/1

Solving for x, the result is 124.

Look at the following table:

Month	*Number of Days in the Month*	*Subtotal of Days*	*Subtotal of Days for a Leap Year*
Jan	31	31	31
Feb	28	59	60
Mar	31	90	91
April	30	120	121
May	31	151	152
June	30	181	182
July	31	212	213
Aug	31	243	244
Sep	30	273	274
Oct	31	304	305
Nov	30	334	335
Dec	31	365	366

As x = 124 so 124 minus 120 for April equals 4 for May, thus the actual date on which the primary direction occurs is May 4, 2013.

Chapter 5

THE PRIMARY DIRECTIONS OF PERSON BB

First, create the chart of Person BB, whose birthdata is July 24, 2000, 2:30 am, Houston TX. We are using Solar Fire to create this chart, which is shown on the next page.

Person BB. Natal Chart. Jul 24, 2000.2:30 AM, CDT +05:00:00, Houston TX, USA,Geocentric, Western Sidereal (Fagan-Allen), Campanus, True Node.

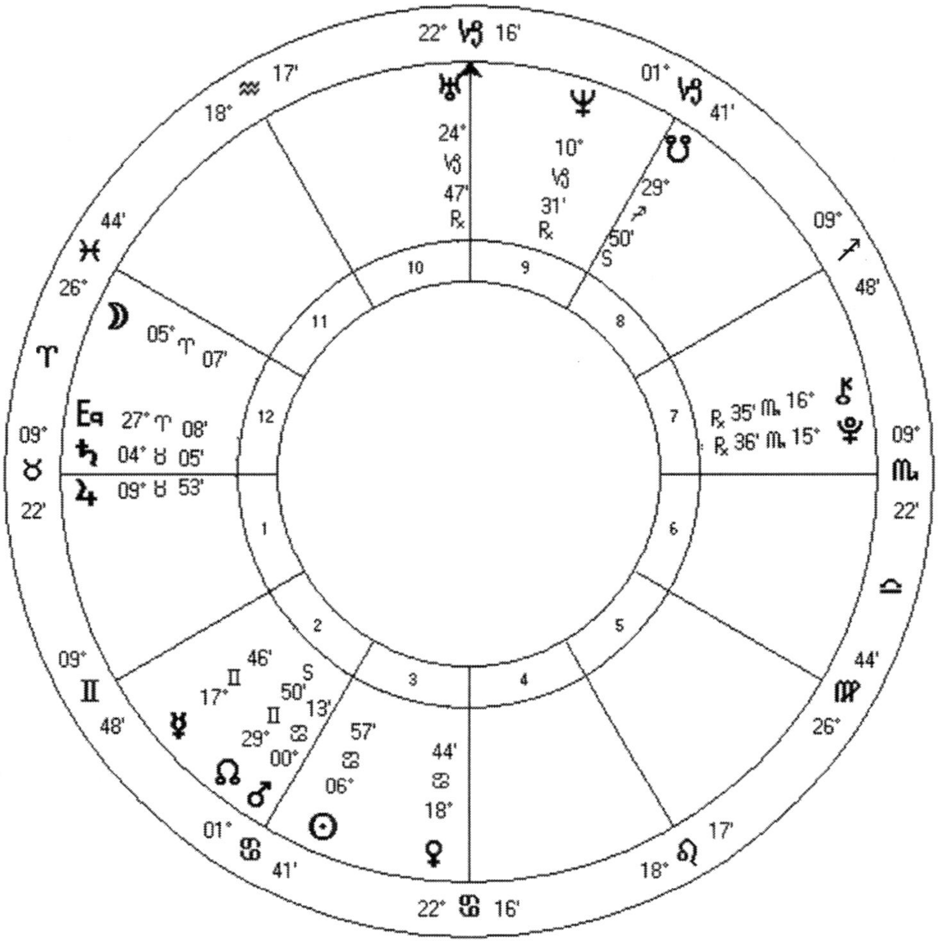

Then cast the chart for the birthdata of Person BB using Janus. After you cast the chart and it is displayed, you must create the astrological speculum of Person BB. Go to the Calculate dropdown and select Diurnal Events (Parans), then from the Settings dropdown, select Options, then select Planet on Angle and hit OK. The following speculum displays.

Person BB. Speculum. Jul 24, 2000. 2:30 AM, CDT +05:00:00, Houston TX, USA, Geocentric, Western Sidereal (Fagan-Bradley) Campanus, True Node.

RAMC 319 28' 07" LST 21:17:52

Diurnal Events: Planet on Angle

Sidereal Time

Planet/Star	Rise	Mc	Set	Ic
Sun	1:28:20	8:15:48	15:03:16	20:15:48
Moon	19:43:42	1:58:37	8:13:32	13:58:37
Mercury	0:05:14	6:53:20	13:41:26	18:53:20
Venus	2:22:26	9:05:32	15:48:38	21:05:32
Mars	0:54:40	7:48:25	14:42:11	19:48:25
Jupiter	21:22:44	4:11:24	11:00:03	16:11:24
Saturn	21:05:49	3:48:21	10:30:52	15:48:21
Uranus	16:05:41	21:28:48	2:51:55	9:28:48
Neptune	15:15:05	20:30:19	1:45:34	8:30:19
Pluto	11:07:00	16:41:26	22:15:52	4:41:26
Moon's North Node	0:54:44	7:46:00	14:37:17	19:46:00
Aldebaran	21:59:33	4:38:49	11:18:06	16:38:49

Now to begin.

First, we convert the date and the year of Person BB's birth to decimal form.

For July 24th, there are 7 days until the end of July (to and including July 31).

To 7 add the 31 days of August, 30 days of September, the 31 days of October, the 30 days of November and the 31 days of December. The total is 160 days left in the year from July 24 up to and including December 31.

Divide 160 by 366 (because 2000 is a leap year.

160/366 equals .44

Make .44 negative by hitting the plus/minus key on the calculator, so you have -.44

Then add 1.000 to -.44 to get .56

We now have the birthday in decimal form:

2000.56

1. Calculate the number of years and days from the birth of Person BB for the Ascendant to Jupiter primary direction to occur.

Take the sidereal time value from the speculum when the Ascendant is directed to Jupiter, by looking at the row of Jupiter and then down the Ascendant (Rise) column. The sidereal time is 21h22m44s. Also note the LST of the birth you get at the top of the speculum, which is 21h17m52s.

Type 21 hit the degrees button, type 22 hit the degrees button, type 44

Hit the subtract button.

Type 21 hit the degrees button, type 17 hit the degrees button, type 52

Hit the equal button to get 4m52s.

By hand, change 4m52s to minutes and seconds. This equals 4m52s.

Change 4m52s to a decimal and divide by 4 by doing the following:

Type 4 hit the degrees button, type 52

Hit the Shift button located in the upper left corner of the calculator.

Then hit the degrees button to get 4.86 which is the decimal.

Hit the divide button, type 4 and hit the equal button.

You will get 1.2.

This is the age at which Jupiter will tend to manifest greatly in Person BB's life.

Now to get the date on which this occurs, we add the birthday in decimal form to the date of the primary direction in decimal form:

2000.56 plus 1.2 equals 2001.76

This is the date on which the primary direction occurs in decimal form.

We now convert the decimal form to the actual date.

x/366 = .76/1

Solving for x, the result is 278.

Look at the following table:

Month	*Number of Days in the Month*	*Subtotal of Days*	*Subtotal of Days for a Leap Year*
Jan	31	31	31
Feb	28	59	60
Mar	31	90	91
April	30	120	121
May	31	151	152
June	30	181	182
July	31	212	213
Aug	31	243	244
Sep	30	273	274
Oct	31	304	305
Nov	30	334	335
Dec	31	365	366

As x = 278 so 278 minus 274 for Sep equals 4 for Oct, thus the actual date on which the primary direction occurs is Oct. 4, 2001.

2. Calculate the number of years and days from the birth of Person BB for the Midheaven to Uranus primary direction to occur.

Take the sidereal time value from the speculum when the Midheaven is directed to Uranus, by looking at the row of Uranus and then down the Midheaven (Mc) column. The sidereal time is 21h28m48s. Also note the LST of the birth you get at the top of the speculum, which is 21h17m52s.

Type 21 hit the degrees button, type 28 hit the degrees button, type 48

Hit the subtract button.

Type 21 hit the degrees button, type 17 hit the degrees button, type 52

Hit the equal button to get 10m56s.

By hand, change 10m56s to minutes and seconds. This equals 10m56s.

Change 10m56s to a decimal and divide by 4 by doing the following:

Type 10 hit the degrees button, type 56

Hit the Shift button located in the upper left corner of the calculator.

Then hit the degrees button to get 10.9333 which is the decimal.

Hit the divide button, type 4 and hit the equal button.

You will get 2.7.

This is the age at which Uranus will tend to manifest greatly in Person BB's life.

Now to get the date on which this occurs, we add the birthday in decimal form to the date of the primary direction in decimal form:

2000.56 plus 2.7 equals 2003.26

This is the date on which the primary direction occurs in decimal form.

We now convert the decimal form to the actual date.

x/366 = .26/1

Solving for x, the result is 95.

Look at the following table:

Month	*Number of Days in the Month*	*Subtotal of Days*	*Subtotal of Days for a Leap Year*
Jan	31	31	31
Feb	28	59	60
Mar	31	90	91
April	30	120	121
May	31	151	152
June	30	181	182
July	31	212	213
Aug	31	243	244
Sep	30	273	274
Oct	31	304	305
Nov	30	334	335
Dec	31	365	366

As x = 95 so 95 minus 91 for March equals 4 for April, thus the actual date on which the primary direction occurs is April 4, 2003.

3. Calculate the number of years and days from the birth of Person BB for the Descendant to Pluto primary direction to occur.

Take the sidereal time value from the speculum when the Descendant is directed to Pluto, by looking at the row of Pluto and then down the Descendant (Set) column. The sidereal time is 22h15m52s. Also note the LST of the birth you get at the top of the speculum, which is 21h17m52s.

Type 22 hit the degrees button [° ′ ″], type 15 hit the degrees button [° ′ ″], type 52

Hit the [−] subtract button.

Type 21 hit the degrees button [° ′ ″], type 17 hit the degrees button [° ′ ″], type 52

Hit the equal button [=] to get 58m.

By hand, change 58m to minutes and seconds. This equals 58m.

Change 58m to a decimal and divide by 4 by doing the following:

Type 58 hit the degrees button , type 0

Hit the Shift button 

located in the upper left corner of the calculator.

Then hit the degrees button  to get 58 which is the decimal.

Hit the divide button 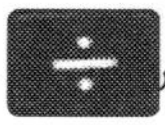, type 4 and hit the equal button .

You will get 14.5.

This is the age at which Pluto will tend to manifest greatly in Person BB's life.

Now to get the date on which this occurs, we add the birthday in decimal form to the date of the primary direction in decimal form:

2000.56 plus 14.5 equals 2015.06

This is the date on which the primary direction occurs in decimal form.

We now convert the decimal form to the actual date.

$x/366 = .06/1$

Solving for x, the result is 22.

Look at the following table:

Month	*Number of Days in the Month*	*Subtotal of Days*	*Subtotal of Days for a Leap Year*
Jan	31	31	31
Feb	28	59	60
Mar	31	90	91
April	30	120	121
May	31	151	152
June	30	181	182
July	31	212	213
Aug	31	243	244
Sep	30	273	274
Oct	31	304	305
Nov	30	334	335
Dec	31	365	366

As x = 22 so 22 minus zero is 22, thus the actual date on which the primary direction occurs is Jan 22, 1991.

4. Calculate the number of years and days from the birth of Person BB for the Ascendant to Mercury primary direction to occur.

Take the sidereal time value from the speculum when the Ascendant is directed to Mercury, by looking at the row of Mercury and then down the Ascendant (Rise) column. The sidereal time is 0h5m14s. Also note the LST of the birth you get at the top of the speculum, which is 21h17m52s.

Because the LST is greater than the sidereal time, add 24h0m0s to 0hr5m14s to get 24h5m14s.

Type 24 hit the degrees button, type 5 hit the degrees button, type 14

Hit the subtract button.

Type 21 hit the degrees button, type 17 hit the degrees button, type 52

Hit the equal button to get 2h47m22s.

By hand, change 2h47m22s to minutes and seconds. This equals 167m22s.

Change 167m22s to a decimal and divide by 4 by doing the following:

Type 167 hit the degrees button, type 22

Hit the Shift button located in the upper left corner of the calculator.

Then hit the degrees button to get 167.3666 which is the decimal.

Hit the divide button, type 4 and hit the equal button.

You will get 41.84.

This is the age at which Mercury will tend to manifest greatly in Person BB's life.

Now to get the date on which this occurs, we add the birthday in decimal form to the date of the primary direction in decimal form:

2000.56 plus 41.84 equals 2042.24

This is the date on which the primary direction occurs in decimal form.

We now convert the decimal form to the actual date.

x/366 = .24/1

Solving for x, the result is 88.

Look at the following table:

Month	*Number of Days in the Month*	*Subtotal of Days*	*Subtotal of Days for a Leap Year*
Jan	31	31	31
Feb	28	59	60
Mar	31	90	91
April	30	120	121
May	31	151	152
June	30	181	182
July	31	212	213
Aug	31	243	244
Sep	30	273	274
Oct	31	304	305
Nov	30	334	335
Dec	31	365	366

As x = 88 so 88 minus 60 for Feb equals 28 for March, thus the actual date on which the primary direction occurs is March 28, 2042.

5. Calculate the number of years and days from the birth of Person BB for the Ascendant to Mars primary direction to occur.

Take the sidereal time value from the speculum when the Ascendant is directed to Mars, by looking at the row of Mars and then down the Ascendant (Rise) column. The sidereal time is 0h54m140s. Also note the LST of the birth you get at the top of the speculum, which is 21h17m52s.

Because the LST is greater than the sidereal time, add 24h0m0s to 0h54m40s to get 24h54m40s.

Type 24 hit the degrees button , type 54 hit the degrees button , type 40

Hit the 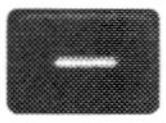 subtract button.

Type 21 hit the degrees button , type 17 hit the degrees button , type 52

Hit the equal button  to get 3h36m48s.

By hand, change 3h36m48s to minutes and seconds. This equals 216m48s.

Change 216m48s to a decimal and divide by 4 by doing the following:

Type 216 hit the degrees button , type 48

Hit the Shift button  located in the upper left corner of the calculator.

Then hit the degrees button  to get 216.8 which is the decimal.

Hit the divide button , type 4 and hit the equal button 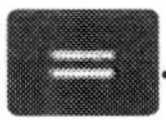.

You will get 54.2.

This is the age at which Mars will tend to manifest greatly in Person BB's life.

Now to get the date on which this occurs, we add the birthday in decimal form to the date of the primary direction in decimal form:

2000.56 plus 54.2 equals 2054.76

This is the date on which the primary direction occurs in decimal form.

We now convert the decimal form to the actual date.

x/366 = .76/1

Solving for x, the result is 278.

Look at the following table:

Month	*Number of Days in the Month*	*Subtotal of Days*	*Subtotal of Days for a Leap Year*
Jan	31	31	31
Feb	28	59	60
Mar	31	90	91
April	30	120	121
May	31	151	152
June	30	181	182
July	31	212	213
Aug	31	243	244
Sep	30	273	274
Oct	31	304	305
Nov	30	334	335
Dec	31	365	366

As x = 278 so 278 minus 274 for Sep equals 4 for October, thus the actual date on which the primary direction occurs is Oct. 4, 2054.

6. Calculate the number of years and days from the birth of Person BB for the Ascendant to the Sun primary direction to occur.

Take the sidereal time value from the speculum when the Ascendant is directed to the Sun, by looking at the row of the Sun and then down the Ascendant (Rise) column. The sidereal time is 1h28m20s. Also note the LST of the birth you get at the top of the speculum, which is 21h17m52s.

Because the LST is greater than the sidereal time, add 24h0m0s to 1h28m20s to get 24h28m20s.

Type 24 hit the degrees button, type 28 hit the degrees button, type 20

Hit the subtract button.

Type 21 hit the degrees button, type 17 hit the degrees button, type 52

Hit the equal button to get 4h10m20s.

By hand, change 4h10m20s to minutes and seconds. This equals 250m20s.

Change 250m20s to a decimal and divide by 4 by doing the following:

Type 250 hit the degrees button, type 20

Hit the Shift button located in the upper left corner of the calculator.

Then hit the degrees button to get 250.33 which is the decimal.

Hit the divide button, type 4 and hit the equal button.

You will get 63.

This is the age at which the Sun will tend to manifest greatly in Person BB's life.

Now to get the date on which this occurs, we add the birthday in decimal form to the date of the primary direction in decimal form:

2000.56 plus 63 equals 2063.56

This is the date on which the primary direction occurs in decimal form.

We now convert the decimal form to the actual date.

x/366 = .56/1

Solving for x, the result is 205.

Look at the following table:

Month	*Number of Days in the Month*	*Subtotal of Days*	*Subtotal of Days for a Leap Year*
Jan	31	31	31
Feb	28	59	60
Mar	31	90	91
April	30	120	121
May	31	151	152
June	30	181	182
July	31	212	213
Aug	31	243	244
Sep	30	273	274
Oct	31	304	305
Nov	30	334	335
Dec	31	365	366

As x = 205 so 205 minus 182 for Jun equals 23 for July, thus the actual date on which the primary direction occurs is July 23, 2063.

7. Calculate the number of years and days from the birth of Person BB for the Descendant to Neptune primary direction to occur.

Take the sidereal time value from the speculum when the Descendant is directed to Neptune, by looking at the row of Neptune and then down the Descendant (Set) column. The sidereal time is 1h35m44s. Also note the LST of the birth you get at the top of the speculum, which is 21h17m52s.

Because the LST is greater than the sidereal time, add 24h0m0s to 1h35m44s to get 25h35m44s.

Type 25 hit the degrees button [° ′ ″], type 35 hit the degrees button [° ′ ″], type 44

Hit the [−] subtract button.

Type 21 hit the degrees button [° ′ ″], type 17 hit the degrees button [° ′ ″], type 52

Hit the equal button [=] to get 4h17m52s.

By hand, change 4h17m52s to minutes and seconds. This equals 257m52s.
Change 257m52s to a decimal and divide by 4 by doing the following:
Type 257 hit the degrees button , type 52
Hit the Shift button  located in the upper left corner of the calculator.
Then hit the degrees button  to get 257.86 which is the decimal.
Hit the divide button 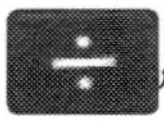, type 4 and hit the equal button .
You will get 64.47.
This is the age at which Neptune will tend to manifest greatly in Person BB's life.

Now to get the date on which this occurs, we add the birthday in decimal form to the date of the primary direction in decimal form:
2000.56 plus 64.47 equals 2065.027
This is the date on which the primary direction occurs in decimal form.
We now convert the decimal form to the actual date.
x/366 = .027/1
Solving for x, the result is 10.
Look at the following table:

Month	*Number of Days in the Month*	*Subtotal of Days*	*Subtotal of Days for a Leap Year*
Jan	31	31	31
Feb	28	59	60
Mar	31	90	91
April	30	120	121
May	31	151	152
June	30	181	182
July	31	212	213
Aug	31	243	244
Sep	30	273	274
Oct	31	304	305
Nov	30	334	335
Dec	31	365	366

As x = 205 so 205 minus 182 for Jun equals 23 for July, thus the actual date on which the primary direction occurs is July 23, 2070.

8. Calculate the number of years and days from the birth of Person BB for the Midheaven to Moon primary direction to occur.

Take the sidereal time value from the speculum when the Midheaven is directed to the Moon, by looking at the row of the Moon and then down the Midheaven (Mc) column. The sidereal time is 1h58m37s. Also note the LST of the birth you get at the top of the speculum, which is 21h17m52s.

Because the LST is greater than the sidereal time, add 24h0m0s to 1h58m37s to get 25h58m37s.

Type 25 hit the degrees button , type 58 hit the degrees button , type 37

Hit the subtract button.

Type 21 hit the degrees button , type 17 hit the degrees button , type 52

Hit the equal button to get 4h40m45s.

By hand, change 4h40m45s to minutes and seconds. This equals 280m45s.

Change 280m45s to a decimal and divide by 4 by doing the following:

Type 280 hit the degrees button , type 45

Hit the Shift button located in the upper left corner of the calculator.

Then hit the degrees button to get 280.75 which is the decimal.

Hit the divide button , type 4 and hit the equal button .

You will get 70.

This is the age at which the Moon will tend to manifest greatly in Person BB's life.

Now to get the date on which this occurs, we add the birthday in decimal form to the date of the primary direction in decimal form:

2000.56 plus 70 equals 2070.56

This is the date on which the primary direction occurs in decimal form.

We now convert the decimal form to the actual date.

x/366 = .56/1

Solving for x, the result is 205.

Look at the following table:

Month	*Number of Days in the Month*	*Subtotal of Days*	*Subtotal of Days for a Leap Year*
Jan	31	31	31
Feb	28	59	60
Mar	31	90	91
April	30	120	121
May	31	151	152
June	30	181	182
July	31	212	213
Aug	31	243	244
Sep	30	273	274
Oct	31	304	305
Nov	30	334	335
Dec	31	365	366

As x = 205 so 205 minus 182 for Jun equals 23 for July, thus the actual date on which the primary direction occurs is July 23, 2070.

9. Calculate the number of years and days from the birth of Person BB for the Ascendant to Venus primary direction to occur.

Take the sidereal time value from the speculum when the Ascendant is directed to Venus, by looking at the row of Venus and then down the Ascendant (Rise) column. The sidereal time is 2h22m26s. Also note the LST of the birth you get at the top of the speculum, which is 21h17m52s.

Because the LST is greater than the sidereal time, add 24h0m0s to 2h22m26s to get 26h22m26s.

Type 26 hit the degrees button [° ' "], type 22 hit the degrees button [° ' "], type 26

Hit the [−] subtract button.

Type 21 hit the degrees button [° ' "], type 17 hit the degrees button [° ' "], type 52

Hit the equal button [=] to get 5h4m34s.

By hand, change 5h4m26s to minutes and seconds. This equals 304m34s.

Change 304m34s to a decimal and divide by 4 by doing the following:

Type 304 hit the degrees button , type 34

Hit the Shift button  located in the upper left corner of the calculator.

Then hit the degrees button  to get 304.566 which is the decimal.

Hit the divide button , type 4 and hit the equal button .

You will get 76.

This is the age at which Venus will tend to manifest greatly in Person BB's life.

Now to get the date on which this occurs, we add the birthday in decimal form to the date of the primary direction in decimal form:

2000.56 plus 76 equals 2076.56

This is the date on which the primary direction occurs in decimal form.

We now convert the decimal form to the actual date.

x/366 = .56/1

Solving for x, the result is 205.

Look at the following table:

Month	*Number of Days in the Month*	*Subtotal of Days*	*Subtotal of Days for a Leap Year*
Jan	31	31	31
Feb	28	59	60
Mar	31	90	91
April	30	120	121
May	31	151	152
June	30	181	182
July	31	212	213
Aug	31	243	244
Sep	30	273	274
Oct	31	304	305
Nov	30	334	335
Dec	31	365	366

As x = 205 so 205 minus 182 for Jun equals 23 for July, thus the actual date on which the primary direction occurs is July 23, 2070.

10. Calculate the number of years and days from the birth of Person BB for the Midheaven to Saturn primary direction to occur.

Take the sidereal time value from the speculum when the Midheaven is directed to Saturn, by looking at the row of Saturn and then down the Midheaven (Mc) column. The sidereal time is 3h48m21s. Also note the LST of the birth you get at the top of the speculum, which is 21h17m52s.

Because the LST is greater than the sidereal time, add 24h0m0s to 3h48m21s to get 27h48m21s.

Type 27 hit the degrees button , type 48 hit the degrees button , type 21

Hit the subtract button.

Type 21 hit the degrees button , type 17 hit the degrees button , type 52

Hit the equal button to get 6h30m29s.

By hand, change 6h30m29s to minutes and seconds. This equals 390m29s.

Change 390m29s to a decimal and divide by 4 by doing the following:

Type 390 hit the degrees button , type 29

Hit the Shift button located in the upper left corner of the calculator.

Then hit the degrees button to get 390.48 which is the decimal.

Hit the divide button , type 4 and hit the equal button .

You will get 97.62.

This is the age at which Saturn will tend to manifest greatly in Person BB's life.

Now to get the date on which this occurs, we add the birthday in decimal form to the date of the primary direction in decimal form:

2000.56 plus 97.62 equals 2098.18

This is the date on which the primary direction occurs in decimal form.

We now convert the decimal form to the actual date.

x/366 = .18/1

Solving for x, the result is 66.

Look at the following table:

Month	*Number of Days in the Month*	*Subtotal of Days*	*Subtotal of Days for a Leap Year*
Jan	31	31	31
Feb	28	59	60
Mar	31	90	91
April	30	120	121
May	31	151	152
June	30	181	182
July	31	212	213
Aug	31	243	244
Sep	30	273	274
Oct	31	304	305
Nov	30	334	335
Dec	31	365	366

As $x = 66$ so 66 minus 60 for Feb equals 6 for March, thus the actual date on which the primary direction occurs is March 6, 2098.

Chapter 6

THE PRIMARY DIRECTIONS OF PERSON CC

First, create the chart of Person CC, whose birthdata is Jan 14, 2012, 1:30 am, Charleston SC. We are using Solar Fire to create this chart, which is shown on the next page.

Person CC. Natal Chart. Jan 14, 2012. 1:30 AM, EST +05:00:00, Charleston SC, USA,Geocentric, Western Sidereal (Fagan-Allen), Campanus, True Node.

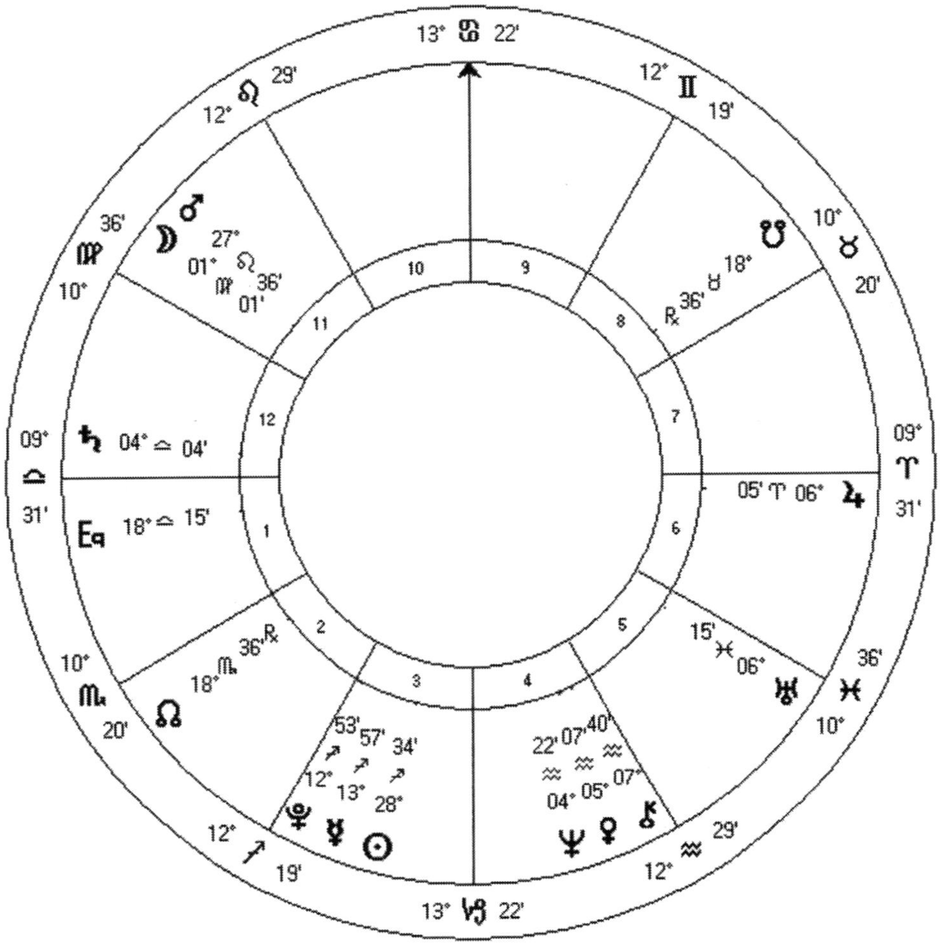

Then cast the chart for the birthdata of Person CC using Janus. After you cast the chart and it is displayed, you must create the astrological speculum of Person CC. Go to the Calculate dropdown and select Diurnal Events (Parans), then from the Settings dropdown, select Options, then select Planet on Angle and hit OK. The following speculum displays.

Person CC. Speculum. Jan 14, 2012. 1:30 AM, EST +05:00:00, Charleston SC, USA, Geocentric, Western Sidereal (Fagan-Bradley) Campanus, True Node.

RAMC 130 42' 57" LST 8:42:51

Diurnal Events: Planet on Angle
Sidereal Time

Planet/Star	Rise	Mc	Set	Ic
Sun	14:39:47	19:41:20	0:42:53	7:41:20
Moon	5:44:49	11:37:02	17:29:15	23:37:02
Mercury	13:45:05	18:38:47	23:32:30	6:38:47
Venus	16:44:55	22:10:49	3:36:43	10:10:49
Mars	5:22:13	11:37:47	17:53:21	23:37:47
Jupiter	19:29:06	1:57:09	8:25:11	13:57:09
Saturn	8:14:13	13:51:20	19:28:28	1:51:20
Uranus	18:05:54	0:05:24	6:04:54	12:05:24
Neptune	16:38:30	22:06:24	3:34:18	10:06:24
Pluto	13:25:08	18:32:58	23:40:48	6:32:58
Moon's North Node	11:50:08	16:48:30	21:46:53	4:48:30
Aldebaran	21:55:08	4:39:29	11:23:50	16:39:29

Now to begin.

First, we convert the date and the year of Person CC's birth to decimal form.

For Jan 14th, there are 17 days until the end of Jan (to and including Jan 31).

To 17 add the 29 days of Feb, 31 days of March, 30 days of April, 31 days of May, 30 days of June, 31 days of July, 31 days of August, 30 days of September, the 31 days of October, the 30 days of November and the 31 days of December. The total is 353 days left in the year from and including Jan 14 up to and including December 31.

Divide 352 by 366 (because 2012 is a leap year).

352/366 equals .96

Make .96 negative by hitting the plus/minus key on the calculator, so you have -.96

Then add 1.000 to -.96 to get .04

We now have the birthday in decimal form:

2012.04

1. Calculate the number of years and days from the birth of Person CC for the IC to Neptune primary direction to occur.

Take the sidereal time value from the speculum when the IC is directed to Neptune, by looking at the row of Neptune and then down the IC column. The sidereal time is 10h06m24s. Also note the LST of the birth you get at the top of the speculum, which is 8h42m51s.

Type 10 hit the degrees button, type 6 hit the degrees button, type 24

Hit the subtract button.

Type 8 hit the degrees button, type 42 hit the degrees button, type 51

Hit the equal button to get 1h23m33s.

By hand, change 1h23m33s to minutes and seconds. This equals 83m33s.

Change 83m33s to a decimal and divide by 4 by doing the following:

Type 83 hit the degrees button, type 33

Hit the Shift button located in the upper left corner of the calculator.

Then hit the degrees button to get 83.55 which is the decimal.

Hit the divide button, type 4 and hit the equal button.

You will get 20.8875.

This is the age at which Neptune will tend to manifest greatly in Person CC's life.

Now to get the date on which this occurs, we add the birthday in decimal form to the date of the primary direction in decimal form:

2012.04 plus 20.8875 equals 2032.9275

This is the date on which the primary direction occurs in decimal form.

We now convert the decimal form to the actual date.

x/366 = .9275/1

Solving for x, the result is 339.

Look at the following table:

Month	*Number of Days in the Month*	*Subtotal of Days*	*Subtotal of Days for a Leap Year*
Jan	31	31	31
Feb	28	59	60
Mar	31	90	91
April	30	120	121
May	31	151	152
June	30	181	182
July	31	212	213
Aug	31	243	244
Sep	30	273	274
Oct	31	304	305
Nov	30	334	335
Dec	31	365	366

As x = 339 so 339 minus 335 for Nov equals 4 for Dec, thus the actual date on which the primary direction occurs is Dec. 4, 2032.

2. Calculate the number of years and days from the birth of Person CC for the IC to Venus primary direction to occur.

Take the sidereal time value from the speculum when the IC is directed to Venus, by looking at the row of Venus and then down the IC column. The sidereal time is 10h10m49s. Also note the LST of the birth you get at the top of the speculum, which is 8h42m51s.

Type 10 hit the degrees button [° ' ''], type 10 hit the degrees button [° ' ''], type 49

Hit the [−] subtract button.

Type 8 hit the degrees button [° ' ''], type 42 hit the degrees button [° ' ''], type 51

Hit the equal button [=] to get 1h27m58s.

By hand, change 1h27m58s to minutes and seconds. This equals 87m58s.

Change 87m58s to a decimal and divide by 4 by doing the following:

Type 87 hit the degrees button [° ' ''], type 58

Hit the Shift button [SHIFT] located in the upper left corner of the calculator.

Then hit the degrees button [° ' ''] to get 87.96 which is the decimal.

Hit the divide button [÷], type 4 and hit the equal button [=].

You will get 22.

This is the age at which Venus will tend to manifest greatly in Person CC's life.

Now to get the date on which this occurs, we add the birthday in decimal form to the date of the primary direction in decimal form:

2012.04 plus 22 equals 2034.04

This is the date on which the primary direction occurs in decimal form.

We now convert the decimal form to the actual date.

x/366 = .04/1

Solving for x, the result is 15.

Look at the following table:

Month	*Number of Days in the Month*	*Subtotal of Days*	*Subtotal of Days for a Leap Year*
Jan	31	31	31
Feb	28	59	60
Mar	31	90	91
April	30	120	121
May	31	151	152
June	30	181	182
July	31	212	213
Aug	31	243	244
Sep	30	273	274
Oct	31	304	305
Nov	30	334	335
Dec	31	365	366

As x = 15 so 15 minus zero for Jan equals 15 for Jan, thus the actual date on which the primary direction occurs is Jan.15, 2034.

3. Calculate the number of years and days from the birth of Person CC for the Midheaven to Moon primary direction to occur.

Take the sidereal time value from the speculum when the Midheaven is directed to the Moon, by looking at the row of the Moon and then down the Midheaven (Mc) column. The sidereal time is 11h37m02s. Also note the LST of the birth you get at the top of the speculum, which is 8h42m51s.

Type 11 hit the degrees button [° ' ''], type 37 hit the degrees button [° ' ''], type 51

Hit the [−] subtract button.

Type 8 hit the degrees button [° ' ''], type 42 hit the degrees button [° ' ''], type 51

Hit the equal button [=] to get 2h54m11s.

By hand, change 2h54m11s to minutes and seconds. This equals 174m11s.

Change 174m11s to a decimal and divide by 4 by doing the following:

Type 174 hit the degrees button , type 11
Hit the Shift button 

located in the upper left corner of the calculator.
Then hit the degrees button  to get 174.183 which is the decimal.
Hit the divide button 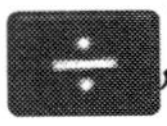, type 4 and hit the equal button .
You will get 43.5458.
This is the age at which the Moon will tend to manifest greatly in Person CC's life.

Now to get the date on which this occurs, we add the birthday in decimal form to the date of the primary direction in decimal form:

2012.04 plus 43.5458 equals 2055.58
This is the date on which the primary direction occurs in decimal form.
We now convert the decimal form to the actual date.
x/366 = .58/1
Solving for x, the result is 212.
Look at the following table:

Month	*Number of Days in the Month*	*Subtotal of Days*	*Subtotal of Days for a Leap Year*
Jan	31	31	31
Feb	28	59	60
Mar	31	90	91
April	30	120	121
May	31	151	152
June	30	181	182
July	31	212	213
Aug	31	243	244
Sep	30	273	274
Oct	31	304	305
Nov	30	334	335
Dec	31	365	366

As x = 212 so 212 minus 182 for Jun equals 30 for Jul, thus the actual date on which the primary direction occurs is July 3, 2055.

4. Calculate the number of years and days from the birth of Person CC for the Midheaven to Mars primary direction to occur.

Take the sidereal time value from the speculum when the Midheaven is directed to Mars, by looking at the row of Mars and then down the Midheaven (Mc) column. The sidereal time is 11h37m47s. Also note the LST of the birth you get at the top of the speculum, which is 8h42m51s.

Type 11 hit the degrees button, type 37 hit the degrees button, type 47

Hit the subtract button.

Type 8 hit the degrees button, type 42 hit the degrees button, type 51

Hit the equal button to get 2h54m56s.

By hand, change 2h54m56s to minutes and seconds. This equals 174m56s.

Change 174m56s to a decimal and divide by 4 by doing the following:

Type 174 hit the degrees button, type 56

Hit the Shift button located in the upper left corner of the calculator.

Then hit the degrees button to get 175 which is the decimal.

Hit the divide button, type 4 and hit the equal button.

You will get 43.733.

This is the age at which Mars will tend to manifest greatly in Person CC's life.

Now to get the date on which this occurs, we add the birthday in decimal form to the date of the primary direction in decimal form:

2012.04 plus 43.733 equals 2055.77

This is the date on which the primary direction occurs in decimal form.

We now convert the decimal form to the actual date.

x/366 = .77/1

Solving for x, the result is 282.

Look at the following table:

Month	*Number of Days in the Month*	*Subtotal of Days*	*Subtotal of Days for a Leap Year*
Jan	31	31	31
Feb	28	59	60
Mar	31	90	91
April	30	120	121
May	31	151	152
June	30	181	182
July	31	212	213
Aug	31	243	244
Sep	30	273	274
Oct	31	304	305
Nov	30	334	335
Dec	31	365	366

As x = 282 so 282 minus 274 for Sep equals 8 for Oct, thus the actual date on which the primary direction occurs is Oct. 8, 2055.

5. Calculate the number of years and days from the birth of Person CC for the IC to Uranus primary direction to occur.

Take the sidereal time value from the speculum when the IC is directed to Uranus, by looking at the row of Uranus and then down the IC column. The sidereal time is 12h05m24s. Also note the LST of the birth you get at the top of the speculum, which is 8h42m51s.

Type 12 hit the degrees button [° ′ ″], type 5 hit the degrees button [° ′ ″], type 24

Hit the [−] subtract button.

Type 8 hit the degrees button [° ′ ″], type 42 hit the degrees button [° ′ ″], type 51

Hit the equal button [=] to get 3h22m33s.

By hand, change 3h22m33s to minutes and seconds. This equals 202m33s.

Change 202m33s to a decimal and divide by 4 by doing the following:

Type 202 hit the degrees button [° ' ''], type 33

Hit the Shift button [SHIFT] located in the upper left corner of the calculator.

Then hit the degrees button [° ' ''] to get 202.55 which is the decimal.

Hit the divide button [÷], type 4 and hit the equal button [=].

You will get 50.64.

This is the age at which Uranus will tend to manifest greatly in Person CC's life.

Now to get the date on which this occurs, we add the birthday in decimal form to the date of the primary direction in decimal form:

2012.04 plus 50.64 equals 2062.68

This is the date on which the primary direction occurs in decimal form.

We now convert the decimal form to the actual date.

x/366 = .68/1

Solving for x, the result is 249.

Look at the following table:

Month	*Number of Days in the Month*	*Subtotal of Days*	*Subtotal of Days for a Leap Year*
Jan	31	31	31
Feb	28	59	60
Mar	31	90	91
April	30	120	121
May	31	151	152
June	30	181	182
July	31	212	213
Aug	31	243	244
Sep	30	273	274
Oct	31	304	305
Nov	30	334	335
Dec	31	365	366

As x = 249 so 249 minus 244 for Aug equals 5 for Sept, thus the actual date on which the primary direction occurs is Sept. 5, 2062.

6. Calculate the number of years and days from the birth of Person CC for the Ascendant to Pluto primary direction to occur.

Take the sidereal time value from the speculum when the Ascendant is directed to Pluto, by looking at the row of Pluto and then down the Ascendant (Rise) column. The sidereal time is 13h25m08s. Also note the LST of the birth you get at the top of the speculum, which is 8h42m51s.

Type 13 hit the degrees button, type 25 hit the degrees button, type 8

Hit the subtract button.

Type 8 hit the degrees button, type 42 hit the degrees button, type 51

Hit the equal button to get 4h42m17s.

By hand, change 4h42m17s to minutes and seconds. This equals 282m17s.

Change 282m17s to a decimal and divide by 4 by doing the following:

Type 282 hit the degrees button, type 17

Hit the Shift button located in the upper left corner of the calculator.

Then hit the degrees button to get 282.28 which is the decimal.

Hit the divide button, type 4 and hit the equal button.

You will get 70.57.

This is the age at which Pluto will tend to manifest greatly in Person CC's life.

Now to get the date on which this occurs, we add the birthday in decimal form to the date of the primary direction in decimal form:

2012.04 plus 70.57 equals 2082.6

This is the date on which the primary direction occurs in decimal form.

We now convert the decimal form to the actual date.

x/366 = .6/1

Solving for x, the result is 220.

Look at the following table:

Month	*Number of Days in the Month*	*Subtotal of Days*	*Subtotal of Days for a Leap Year*
Jan	31	31	31
Feb	28	59	60
Mar	31	90	91
April	30	120	121
May	31	151	152
June	30	181	182
July	31	212	213
Aug	31	243	244
Sep	30	273	274
Oct	31	304	305
Nov	30	334	335
Dec	31	365	366

As x = 220 so 220 minus 213 for Jul equals 7 for Aug, thus the actual date on which the primary direction occurs is Aug. 7, 2082.

7. Calculate the number of years and days from the birth of Person CC for the Ascendant to Mercury primary direction to occur.

Take the sidereal time value from the speculum when the Ascendant is directed to Mercury, by looking at the row of Mercury and then down the Ascendant (Rise) column. The sidereal time is 13h45m05s. Also note the LST of the birth you get at the top of the speculum, which is 8h42m51s.

Type 13 hit the degrees button, type 45 hit the degrees button, type 5

Hit the subtract button.

Type 8 hit the degrees button, type 42 hit the degrees button, type 51

Hit the equal button to get 5h2m14s.

By hand, change 5h2m14s to minutes and seconds. This equals 302m14s.

Change 302m14s to a decimal and divide by 4 by doing the following:

Type 282 hit the degrees button , type 17
Hit the Shift button 

located in the upper left corner of the calculator.
Then hit the degrees button  to get 302.23 which is the decimal.
Hit the divide button , type 4 and hit the equal button .
You will get 75.558.
This is the age at which Mercury will tend to manifest greatly in Person CC's life.

Now to get the date on which this occurs, we add the birthday in decimal form to the date of the primary direction in decimal form:
2012.04 plus 75.558 equals 2087.6
This is the date on which the primary direction occurs in decimal form.
We now convert the decimal form to the actual date.
x/366 = .6/1
Solving for x, the result is 220.
Look at the following table:

Month	*Number of Days in the Month*	*Subtotal of Days*	*Subtotal of Days for a Leap Year*
Jan	31	31	31
Feb	28	59	60
Mar	31	90	91
April	30	120	121
May	31	151	152
June	30	181	182
July	31	212	213
Aug	31	243	244
Sep	30	273	274
Oct	31	304	305
Nov	30	334	335
Dec	31	365	366

As x = 220 so 220 minus 213 for Jul equals 7 for Aug, thus the actual date on which the primary direction occurs is Aug. 7, 2087.

8. Calculate the number of years and days from the birth of Person CC for the Midheaven to Saturn primary direction to occur.

Take the sidereal time value from the speculum when the Midheaven is directed to Saturn, by looking at the row of Saturn and then down the Midheaven (Mc) column. The sidereal time is 13h51m20s. Also note the LST of the birth you get at the top of the speculum, which is 8h42m51s.

Type 13 hit the degrees button, type 51 hit the degrees button, type 20

Hit the subtract button.

Type 8 hit the degrees button, type 42 hit the degrees button, type 51

Hit the equal button to get 5h8m29s.

By hand, change 5h8m29s to minutes and seconds. This equals 308m29s.

Change 308m29s to a decimal and divide by 4 by doing the following:

Type 308 hit the degrees button, type 29

Hit the Shift button located in the upper left corner of the calculator.

Then hit the degrees button to get 308.48 which is the decimal.

Hit the divide button, type 4 and hit the equal button.

You will get 77.12.

This is the age at which Saturn will tend to manifest greatly in Person CC's life.

Now to get the date on which this occurs, we add the birthday in decimal form to the date of the primary direction in decimal form:

2012.04 plus 77.12 equals 2089.16

This is the date on which the primary direction occurs in decimal form.

We now convert the decimal form to the actual date.

x/366 = .16/1

Solving for x, the result is 59.

Look at the following table:

Month	*Number of Days in the Month*	*Subtotal of Days*	*Subtotal of Days for a Leap Year*
Jan	31	31	31
Feb	28	59	60
Mar	31	90	91
April	30	120	121
May	31	151	152
June	30	181	182
July	31	212	213
Aug	31	243	244
Sep	30	273	274
Oct	31	304	305
Nov	30	334	335
Dec	31	365	366

As x = 59 so 59 minus 31 for Jan equals 28 for Feb, thus the actual date on which the primary direction occurs is Feb 28, 2089.

9. Calculate the number of years and days from the birth of Person CC for the IC to Jupiter primary direction to occur.

Take the sidereal time value from the speculum when the IC is directed to Jupiter, by looking at the row of Jupiter and then down the IC column. The sidereal time is 13h57m09s. Also note the LST of the birth you get at the top of the speculum, which is 8h42m51s.

Type 13 hit the degrees button `° ' "`, type 57 hit the degrees button `° ' "`, type 9

Hit the `−` subtract button.

Type 8 hit the degrees button `° ' "`, type 42 hit the degrees button `° ' "`, type 51

Hit the equal button `=` to get 5h14m18s.

By hand, change 5h14m18s to minutes and seconds. This equals 314m18s.

Change 314m18s to a decimal and divide by 4 by doing the following:

Type 314 hit the degrees button `° ' "`, type 18

Hit the Shift button  located in the upper left corner of the calculator.
Then hit the degrees button  to get 314.3 which is the decimal.
Hit the divide button , type 4 and hit the equal button 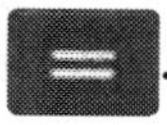.
You will get 77.575.
This is the age at which Jupiter will tend to manifest greatly in Person CC's life.

Now to get the date on which this occurs, we add the birthday in decimal form to the date of the primary direction in decimal form:

2012.04 plus 77.575 equals 2089.615

This is the date on which the primary direction occurs in decimal form.

We now convert the decimal form to the actual date.

x/366 = .615/1

Solving for x, the result is 225.

Look at the following table:

Month	*Number of Days in the Month*	*Subtotal of Days*	*Subtotal of Days for a Leap Year*
Jan	31	31	31
Feb	28	59	60
Mar	31	90	91
April	30	120	121
May	31	151	152
June	30	181	182
July	31	212	213
Aug	31	243	244
Sep	30	273	274
Oct	31	304	305
Nov	30	334	335
Dec	31	365	366

As x = 225 so 225 minus 213 for Jul equals 12 for Aug, thus the actual date on which the primary direction occurs is Aug 12, 2090.

10. Calculate the number of years and days from the birth of Person CC for the Ascendant to Sun primary direction to occur.

Take the sidereal time value from the speculum when the Ascendant is directed to the Sun, by looking at the row of the Sun and then down the Ascendant (Rise) column. The sidereal time is 14h39m47s. Also note the LST of the birth you get at the top of the speculum, which is 8h42m51s.

Type 14 hit the degrees button, type 39 hit the degrees button, type 47

Hit the subtract button.

Type 8 hit the degrees button, type 42 hit the degrees button, type 51

Hit the equal button to get 5h56m56s.

By hand, change 5h56m56s to minutes and seconds. This equals 356m56s.

Change 356m56s to a decimal and divide by 4 by doing the following:

Type 356 hit the degrees button, type 56

Hit the Shift button located in the upper left corner of the calculator.

Then hit the degrees button to get 356.9 which is the decimal.

Hit the divide button, type 4 and hit the equal button.

You will get 89.233.

This is the age at which the Sun will tend to manifest greatly in Person CC's life.

Now to get the date on which this occurs, we add the birthday in decimal form to the date of the primary direction in decimal form:

2012.04 plus 89.233 equals 2101.27

This is the date on which the primary direction occurs in decimal form.

We now convert the decimal form to the actual date.

x/366 = .27/1

Solving for x, the result is 99.

Look at the following table:

Month	*Number of Days in the Month*	*Subtotal of Days*	*Subtotal of Days for a Leap Year*
Jan	31	31	31
Feb	28	59	60
Mar	31	90	91
April	30	120	121
May	31	151	152
June	30	181	182
July	31	212	213
Aug	31	243	244
Sep	30	273	274
Oct	31	304	305
Nov	30	334	335
Dec	31	365	366

As x = 99 so 99 minus 91 for March equals 8 for April thus the actual date on which the primary direction occurs is April 8, 2101.

Chapter 7

THE PRIMARY DIRECTIONS OF PERSON DD

First, create the chart of Person DD, whose birthdata is Oct 12, 1995, 10:30 pm, Los Angeles CA. We are using Solar Fire to create this chart, which is shown on the next page.

Person DD. Natal Chart. Oct 12, 1995. 10:30 PM, PDT +07:00:00, Los Angeles CA, USA,Geocentric, Western Sidereal (Fagan-Allen), Campanus, True Node.

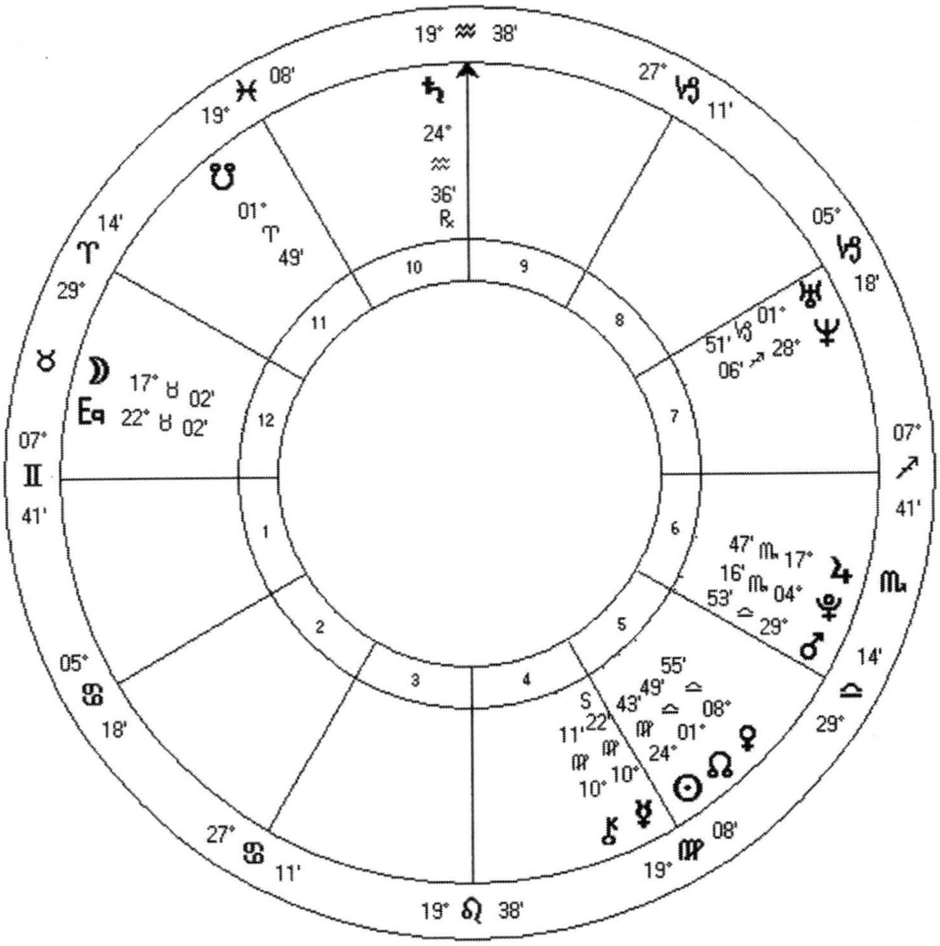

Then cast the chart for the birthdata of Person DD using Janus. After you cast the chart and it is displayed, you must create the astrological speculum of Person DD. Go to the Calculate dropdown and select Diurnal Events (Parans), then from the Settings dropdown, select Options, then select Planet on Angle and hit OK. The following speculum displays.

Person DD. Speculum. Oct 12, 1995. 10:30 PM, PDT +07:00:00, Los Angeles CA, USA, Geocentric, Western Sidereal (Fagan-Bradley) Campanus, True Node.

RAMC 345 34' 19" LST 23:02:17

Diurnal Events: Planet on Angle
Sidereal Time

Planet/Star	Rise	Mc	Set	Ic
Sun	7:32:20	13:11:39	18:50:58	1:11:39
Moon	21:50:47	4:42:57	11:35:07	16:42:57
Mercury	6:23:34	12:19:18	18:15:02	0:19:18
Venus	8:39:51	14:06:20	19:32:49	2:06:20
Mars	10:23:32	15:28:13	20:32:54	3:28:13
Jupiter	11:47:19	16:44:15	21:41:11	4:44:15
Saturn	17:41:40	23:24:19	5:06:57	11:24:19
Uranus	14:56:07	19:54:43	0:53:20	7:54:43
Neptune	14:38:12	19:38:04	0:37:55	7:38:04
Pluto	10:18:00	15:58:17	21:38:33	3:58:17
Moon's North Node	8:06:20	13:38:19	19:10:19	1:38:19
Aldebaran	21:52:04	4:38:33	11:25:02	16:38:33

Now to begin.

First, we convert the date and the year of Person DD's birth to decimal form.

For Oct 12th, there are 19 days until the end of Oct (to and including Oct 31).

To 19 add the 30 days of November and the 31 days of December. The total is 80 days left in the year from Oct 12 up to and including December 31.

Divide 80 by 365.

80/365 equals .219

Make .219 negative by hitting the plus/minus key on the calculator, so you have -.219

Then add 1.000 to -.2219 to get .78

We now have the birthday in decimal form:

1995.78

1. Calculate the number of years and days from the birth of Person DD for the Midheaven to Saturn primary direction to occur.

Take the sidereal time value from the speculum when the Midheaven is directed to Saturn, by looking at the row of Saturn and then down the Midheaven (Mc) column. The sidereal time is 23h24m19s. Also note the LST of the birth you get at the top of the speculum, which is 23h02m17s.

Type 23 hit the degrees button, type 24 hit the degrees button, type 19

Hit the subtract button.

Type 23 hit the degrees button, type 2 hit the degrees button, type 17

Hit the equal button to get 22m02s.

By hand, change 22m02s to minutes and seconds. This equals 22m2s.

Change 22m2s to a decimal and divide by 4 by doing the following:

Type 22 hit the degrees button, type 2

Hit the Shift button located in the upper left corner of the calculator.

Then hit the degrees button to get 22.03 which is the decimal.

Hit the divide button, type 4 and hit the equal button.

You will get 5.5.

This is the age at which Saturn will tend to manifest greatly in Person DD's life.

Now to get the date on which this occurs, we add the birthday in decimal form to the date of the primary direction in decimal form:

1995.78 plus 5.5 equals 2001.28

This is the date on which the primary direction occurs in decimal form.

We now convert the decimal form to the actual date.

x/365 = .28/1

Solving for x, the result is 102.

Look at the following table:

Month	*Number of Days in the Month*	*Subtotal of Days*	*Subtotal of Days for a Leap Year*
Jan	31	31	31
Feb	28	59	60
Mar	31	90	91
April	30	120	121
May	31	151	152
June	30	181	182
July	31	212	213
Aug	31	243	244
Sep	30	273	274
Oct	31	304	305
Nov	30	334	335
Dec	31	365	366

As x = 102 so 102 minus 90 for March equals 12 for April, thus the actual date on which the primary direction occurs is April 12, 2001.

2. Calculate the number of years and days from the birth of Person DD for the IC to Mercury primary direction to occur.

Take the sidereal time value from the speculum when the IC is directed to Mercury, by looking at the row of Mercury and then down the IC column. The sidereal time is 0h19m18s. Also note the LST of the birth you get at the top of the speculum, which is 23h02m17s.

As the sidereal time is less than the LST, add 24h0m0s to the sidereal time. The result is 24h19m18s. Then from 24h19m18s subtract 23h02m17s.

Type 24 hit the degrees button, type 19 hit the degrees button, type 18

Hit the subtract button.

Type 23 hit the degrees button, type 2 hit the degrees button, type 17

Hit the equal button to get 1h17m1s.

By hand, change 1h17m1s to minutes and seconds. This equals 77m1s.

Change 77m1s to a decimal and divide by 4 by doing the following:

Type 77 hit the degrees button, type 1

Hit the Shift button located in the upper left corner of the calculator.

Then hit the degrees button to get 77 which is the decimal.

Hit the divide button, type 4 and hit the equal button.

You will get 19.254.

This is the age at which Mercury will tend to manifest greatly in Person DD's life.

Now to get the date on which this occurs, we add the birthday in decimal form to the date of the primary direction in decimal form:

1995.78 plus 19.254 equals 2015.034

This is the date on which the primary direction occurs in decimal form.

We now convert the decimal form to the actual date.

x/365 = .034/1

Solving for x, the result is 12.

Look at the following table:

Month	*Number of Days in the Month*	*Subtotal of Days*	*Subtotal of Days for a Leap Year*
Jan	31	31	31
Feb	28	59	60
Mar	31	90	91
April	30	120	121
May	31	151	152
June	30	181	182
July	31	212	213
Aug	31	243	244
Sep	30	273	274
Oct	31	304	305
Nov	30	334	335
Dec	31	365	366

As x = 12 so 12 minus zero is 12 for Jan, thus the actual date on which the primary direction occurs is Jan. 12, 2015.

3. Calculate the number of years and days from the birth of Person DD for the Descendant to Neptune primary direction to occur.

Take the sidereal time value from the speculum when the Descendant is directed to Neptune, by looking at the row of Neptune and then down the Descendant (Set) column. The sidereal time is 0h37m55s. Also note the LST of the birth you get at the top of the speculum, which is 23h02m17s.

As the sidereal time is less than the LST, add 24h0m0s to the sidereal time. The result is 24h37m55s. Then from 24h37m55s subtract 23h02m17s.

Type 24 hit the degrees button [° ′ ″], type 37 hit the degrees button [° ′ ″], type 55

Hit the [−] subtract button.

Type 23 hit the degrees button [° ′ ″], type 2 hit the degrees button [° ′ ″], type 17

Hit the equal button [=] to get 1h35m38s.

By hand, change 1h35m38s to minutes and seconds. This equals 95m38s.

Change 95m38s to a decimal and divide by 4 by doing the following:

Type 95 hit the degrees button , type 38
Hit the Shift button  located in the upper left corner of the calculator.
Then hit the degrees button  to get 96.6333 which is the decimal.
Hit the divide button 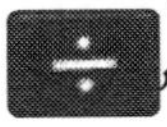, type 4 and hit the equal button .
You will get 23.9.
This is the age at which Neptune will tend to manifest greatly in Person DD's life.

Now to get the date on which this occurs, we add the birthday in decimal form to the date of the primary direction in decimal form:

1995.78 plus 23.9 equals 2019.68

This is the date on which the primary direction occurs in decimal form.

We now convert the decimal form to the actual date.

x/365 = .68/1

Solving for x, the result is 248.

Look at the following table:

Month	*Number of Days in the Month*	*Subtotal of Days*	*Subtotal of Days for a Leap Year*
Jan	31	31	31
Feb	28	59	60
Mar	31	90	91
April	30	120	121
May	31	151	152
June	30	181	182
July	31	212	213
Aug	31	243	244
Sep	30	273	274
Oct	31	304	305
Nov	30	334	335
Dec	31	365	366

As x = 248 so 248 minus 243 for Aug equals 5 for Sep, thus the actual date on which the primary direction occurs is Sep. 5, 2019.

4. Calculate the number of years and days from the birth of Person DD for the Descendant to Uranus primary direction to occur.

Take the sidereal time value from the speculum when the Descendant is directed to Uranus, by looking at the row of Uranus and then down the Descendant (Set) column. The sidereal time is 0h53m20s. Also note the LST of the birth you get at the top of the speculum, which is 23h02m17s.

As the sidereal time is less than the LST, add 24h0m0s to the sidereal time. The result is 24h53m20s. Then from 24h53m20s subtract 23h02m17s.

Type 24 hit the degrees button , type 53 hit the degrees button , type 20

Hit the subtract button.

Type 23 hit the degrees button , type 2 hit the degrees button , type 17

Hit the equal button to get 1h51m3s.

By hand, change 1h51m3s to minutes and seconds. This equals 111m3s.

Change 111m3s to a decimal and divide by 4 by doing the following:

Type 111 hit the degrees button , type 3

Hit the Shift button located in the upper left corner of the calculator.

Then hit the degrees button to get 111.05 which is the decimal.

Hit the divide button , type 4 and hit the equal button .

You will get 27.7625.

This is the age at which Uranus will tend to manifest greatly in Person DD's life.

Now to get the date on which this occurs, we add the birthday in decimal form to the date of the primary direction in decimal form:

1995.78 plus 27.7625 equals 2023.5425

This is the date on which the primary direction occurs in decimal form.

We now convert the decimal form to the actual date.

x/365 = .5425/1

Solving for x, the result is 198.

Look at the following table:

Month	*Number of Days in the Month*	*Subtotal of Days*	*Subtotal of Days for a Leap Year*
Jan	31	31	31
Feb	28	59	60
Mar	31	90	91
April	30	120	121
May	31	151	152
June	30	181	182
July	31	212	213
Aug	31	243	244
Sep	30	273	274
Oct	31	304	305
Nov	30	334	335
Dec	31	365	366

As x = 198 so 198 minus 181 for June equals 17 for July, thus the actual date on which the primary direction occurs is July 17, 2023.

5. Calculate the number of years and days from the birth of Person DD for the IC to Sun primary direction to occur.

Take the sidereal time value from the speculum when the IC is directed to the Sun, by looking at the row of the Sun and then down the IC column. The sidereal time is 1h11m39s. Also note the LST of the birth you get at the top of the speculum, which is 23h02m17s.

As the sidereal time is less than the LST, add 24h0m0s to the sidereal time. The result is 25h11m39s. Then from 25h11m39s subtract 23h02m17s.

Type 25 hit the degrees button ° ' '', type 11 hit the degrees button ° ' '', type 39

Hit the − subtract button.

Type 23 hit the degrees button ° ' '', type 2 hit the degrees button ° ' '', type 17

Hit the equal button = to get 2h9m22s.

By hand, change 2h9m22s to minutes and seconds. This equals 129m22s.

Change 129m22s to a decimal and divide by 4 by doing the following:

Type 129 hit the degrees button , type 22

Hit the Shift button  located in the upper left corner of the calculator.

Then hit the degrees button  to get 129.3666 which is the decimal.

Hit the divide button , type 4 and hit the equal button .

You will get 32.34.

This is the age at which the Sun will tend to manifest greatly in Person DD's life.

Now to get the date on which this occurs, we add the birthday in decimal form to the date of the primary direction in decimal form:

1995.78 plus 32.34 equals 2028.12

This is the date on which the primary direction occurs in decimal form.

We now convert the decimal form to the actual date.

x/365 = .12/1

Solving for x, the result is 44.

Look at the following table:

Month	*Number of Days in the Month*	*Subtotal of Days*	*Subtotal of Days for a Leap Year*
Jan	31	31	31
Feb	28	59	60
Mar	31	90	91
April	30	120	121
May	31	151	152
June	30	181	182
July	31	212	213
Aug	31	243	244
Sep	30	273	274
Oct	31	304	305
Nov	30	334	335
Dec	31	365	366

As x = 44 so 44 minus 31 for Jan equals 13 for Feb, thus the actual date on which the primary direction occurs is Feb 13, 2028.

6. Calculate the number of years and days from the birth of Person DD for the IC to Venus primary direction to occur.

Take the sidereal time value from the speculum when the IC is directed to Venus, by looking at the row of Venus and then down the IC column. The sidereal time is 2h06m20s. Also note the LST of the birth you get at the top of the speculum, which is 23h02m17s.

As the sidereal time is less than the LST, add 24h0m0s to the sidereal time. The result is 26h6m20s. Then from 26h6m20s subtract 23h02m17s.

Type 26 hit the degrees button , type 6 hit the degrees button , type 20

Hit the subtract button.

Type 23 hit the degrees button , type 2 hit the degrees button , type 17

Hit the equal button to get 3h4m3s.

By hand, change 3h4m3s to minutes and seconds. This equals 184m3s.

Change 184m3s to a decimal and divide by 4 by doing the following:

Type 184 hit the degrees button , type 3

Hit the Shift button located in the upper left corner of the calculator.

Then hit the degrees button to get 184.05 which is the decimal.

Hit the divide button , type 4 and hit the equal button .

You will get 46.

This is the age at which the Sun will tend to manifest greatly in Person DD's life.

Now to get the date on which this occurs, we add the birthday in decimal form to the date of the primary direction in decimal form:

1995.78 plus 46 equals 2041.78

This is the date on which the primary direction occurs in decimal form.

We now convert the decimal form to the actual date.

x/365 = .78/1

Solving for x, the result is 285.

Look at the following table:

Month	*Number of Days in the Month*	*Subtotal of Days*	*Subtotal of Days for a Leap Year*
Jan	31	31	31
Feb	28	59	60
Mar	31	90	91
April	30	120	121
May	31	151	152
June	30	181	182
July	31	212	213
Aug	31	243	244
Sep	30	273	274
Oct	31	304	305
Nov	30	334	335
Dec	31	365	366

As x = 285 so 285 minus 273 for Sep equals 12 for Oct, thus the actual date on which the primary direction occurs is Oct 12, 2041.

7. Calculate the number of years and days from the birth of Person DD for the IC to Mars primary direction to occur.

Take the sidereal time value from the speculum when the IC is directed to Mars, by looking at the row of Mars and then down the IC column. The sidereal time is 3h28m13s. Also note the LST of the birth you get at the top of the speculum, which is 23h02m17s.

As the sidereal time is less than the LST, add 24h0m0s to the sidereal time. The result is 27h28m13s. Then from 27h28m13s subtract 23h02m17s.

Type 27 hit the degrees button ° ′ ″, type 28 hit the degrees button ° ′ ″, type 13

Hit the − subtract button.

Type 23 hit the degrees button ° ′ ″, type 2 hit the degrees button ° ′ ″, type 17

Hit the equal button = to get 4h25m56s.

By hand, change 4h25m56s to minutes and seconds. This equals 265m56s.

Change 265m56s to a decimal and divide by 4 by doing the following:

Type 265 hit the degrees button , type 56

Hit the Shift button  located in the upper left corner of the calculator.

Then hit the degrees button  to get 265.93 which is the decimal.

Hit the divide button , type 4 and hit the equal button .

You will get 66.483.

This is the age at which Mars will tend to manifest greatly in Person DD's life.

Now to get the date on which this occurs, we add the birthday in decimal form to the date of the primary direction in decimal form:

1995.78 plus 66.483 equals 2062.26

This is the date on which the primary direction occurs in decimal form.

We now convert the decimal form to the actual date.

x/365 = .26/1

Solving for x, the result is 95.

Look at the following table:

Month	*Number of Days in the Month*	*Subtotal of Days*	*Subtotal of Days for a Leap Year*
Jan	31	31	31
Feb	28	59	60
Mar	31	90	91
April	30	120	121
May	31	151	152
June	30	181	182
July	31	212	213
Aug	31	243	244
Sep	30	273	274
Oct	31	304	305
Nov	30	334	335
Dec	31	365	366

As x = 95 so 95 minus 90 for Mar equals 5 for April, thus the actual date on which the primary direction occurs is April 5, 2062.

8. Calculate the number of years and days from the birth of Person DD for the IC to Pluto primary direction to occur.

Take the sidereal time value from the speculum when the IC is directed to Pluto, by looking at the row of Pluto and then down the IC column. The sidereal time is 3h58m17s. Also note the LST of the birth you get at the top of the speculum, which is 23h02m17s.

As the sidereal time is less than the LST, add 24h0m0s to the sidereal time. The result is 27h58m17s. Then from 27h58m17s subtract 23h02m17s.

Type 27 hit the degrees button, type 58 hit the degrees button, type 17

Hit the subtract button.

Type 23 hit the degrees button, type 2 hit the degrees button, type 17

Hit the equal button to get 4h56m0s.

By hand, change 4h56m0s to minutes and seconds. This equals 296m0s.

Change 296m0s to a decimal and divide by 4 by doing the following:

Type 296 hit the degrees button, type 0

Hit the Shift button located in the upper left corner of the calculator.

Then hit the degrees button to get 296 which is the decimal.

Hit the divide button, type 4 and hit the equal button.

You will get 74.

This is the age at which Pluto will tend to manifest greatly in Person DD's life.

Now to get the date on which this occurs, we add the birthday in decimal form to the date of the primary direction in decimal form:

1995.78 plus 74 equals 2069.78

This is the date on which the primary direction occurs in decimal form.

We now convert the decimal form to the actual date.

x/365 = .78/1

Solving for x, the result is 285.

Look at the following table:

Month	*Number of Days in the Month*	*Subtotal of Days*	*Subtotal of Days for a Leap Year*
Jan	31	31	31
Feb	28	59	60
Mar	31	90	91
April	30	120	121
May	31	151	152
June	30	181	182
July	31	212	213
Aug	31	243	244
Sep	30	273	274
Oct	31	304	305
Nov	30	334	335
Dec	31	365	366

As x = 285 so 285 minus 273 for Sep equals 12 for Oct, thus the actual date on which the primary direction occurs is Oct 12, 2069.

9. Calculate the number of years and days from the birth of Person DD for the Midheaven to the Moon primary direction to occur.

Take the sidereal time value from the speculum when the Midheaven is directed to the Moon, by looking at the row of the Moon and then down the Midheaven (Mc) column. The sidereal time is 4h42m57s. Also note the LST of the birth you get at the top of the speculum, which is 23h02m17s.

As the sidereal time is less than the LST, add 24h0m0s to the sidereal time. The result is 28h42m57s. Then from 28h42m57s subtract 23h02m17s.

Type 28 hit the degrees button ° ′ ″, type 42 hit the degrees button ° ′ ″, type 57

Hit the − subtract button.

Type 23 hit the degrees button ° ′ ″, type 2 hit the degrees button ° ′ ″, type 17

Hit the equal button = to get 5h40m40s.

By hand, change 5h40m40s to minutes and seconds. This equals 340m40s.

Change 340m40s to a decimal and divide by 4 by doing the following:

Type 340 hit the degrees button , type 40
Hit the Shift button  located in the upper left corner of the calculator.
Then hit the degrees button  to get 340.66 which is the decimal.
Hit the divide button , type 4 and hit the equal button .
You will get 85.166.
This is the age at which the Moon will tend to manifest greatly in Person DD's life.

Now to get the date on which this occurs, we add the birthday in decimal form to the date of the primary direction in decimal form:

1995.78 plus 85.166 equals 2080.946

This is the date on which the primary direction occurs in decimal form.

We now convert the decimal form to the actual date.

x/365 = .946/1

Solving for x, the result is 345.

Look at the following table:

Month	*Number of Days in the Month*	*Subtotal of Days*	*Subtotal of Days for a Leap Year*
Jan	31	31	31
Feb	28	59	60
Mar	31	90	91
April	30	120	121
May	31	151	152
June	30	181	182
July	31	212	213
Aug	31	243	244
Sep	30	273	274
Oct	31	304	305
Nov	30	334	335
Dec	31	365	366

As x = 345 so 345 minus 334 for Nov equals 11 for Dec, thus the actual date on which the primary direction occurs is Dec 11, 2080.

10. Calculate the number of years and days from the birth of Person DD for the IC to Jupiter primary direction to occur.

Take the sidereal time value from the speculum when the IC is directed to Jupiter, by looking at the row of Jupiter and then down the IC column. The sidereal time is 4h44m15s. Also note the LST of the birth you get at the top of the speculum, which is 23h02m17s.

As the sidereal time is less than the LST, add 24h0m0s to the sidereal time. The result is 28h44m15s. Then from 28h44m15s subtract 23h02m17s.

Type 28 hit the degrees button [° ' "], type 44 hit the degrees button [° ' "], type 15

Hit the [−] subtract button.

Type 23 hit the degrees button [° ' "], type 2 hit the degrees button [° ' "], type 17

Hit the equal button [=] to get 5h41m58s.

By hand, change 5h41m58s to minutes and seconds. This equals 341m58s.

Change 341m58s to a decimal and divide by 4 by doing the following:

Type 341 hit the degrees button [° ' "], type 58

Hit the Shift button [SHIFT] located in the upper left corner of the calculator.

Then hit the degrees button [° ' "] to get 341.96 which is the decimal.

Hit the divide button [÷], type 4 and hit the equal button [=].

You will get 85.49.

This is the age at which Jupiter will tend to manifest greatly in Person DD's life.

Now to get the date on which this occurs, we add the birthday in decimal form to the date of the primary direction in decimal form:

1995.78 plus 85.49 equals 2081.27

This is the date on which the primary direction occurs in decimal form.

We now convert the decimal form to the actual date.

x/365 = .27/1

Solving for x, the result is 99.

Look at the following table:

Month	*Number of Days in the Month*	*Subtotal of Days*	*Subtotal of Days for a Leap Year*
Jan	31	31	31
Feb	28	59	60
Mar	31	90	91
April	30	120	121
May	31	151	152
June	30	181	182
July	31	212	213
Aug	31	243	244
Sep	30	273	274
Oct	31	304	305
Nov	30	334	335
Dec	31	365	366

As $x = 99$ so 99 minus 90 for March equals 9 for April, thus the actual date on which the primary direction occurs is April 9, 2081.

Chapter 8

PRIMARY DIRECTIONS AND CHART RECTIFICATION

The use of primary directions can also be used to correct your birthtime:

- If all the past major events resulting from all the past primary directions of all the planets occurred a year earlier in your life, subtract 4 minutes from your birthtime to get the more accurate birthtime.
- If all the past major events resulting from all the past primary directions of all the planets occurred a year later in your life, add 4 minutes to your birthtime to get the more accurate birthtime.

Then when you have a more accurate birthtime, have your chart rectified by a rectification expert in astrology to confirm the new birthtime.

Resources

To implement the calculations in this book, consider using the following resources.

- Solar Fire astrological software v9.0 or higher for charts.

  SOLAR FIRE SETTINGS:
  House System: Campanus
  Zodiac: Fagan-Allen
  Coordinate System: Geocentric
  Angle Progressions: Naibod in RA
  Lunar Node: True Mode
  Default Houses (Sidereal): Campanus
  Default Zodiac: Sidereal (Vedic)
  Ayanamsa for Sidereal Charts: Fagan-Allen
  Composite Chart Houses: Anchor on 10th House
  Display Options: Proportional Houses OFF

- Janus astrological software v4.3 or higher (for speculums).

  JANUS SETTINGS:
  House System: Campanus.
  Coordinates: Geocentric.
  Zodiac: Fagan-Bradley.
  Moon's Node: True.

Notes

Notes

Notes

42443596R00062

Made in the USA
Middletown, DE
11 April 2017